FOREWORD

Everyone said that producing a book would be hard work. They were right. The fact that this one is about new technology which combines at least two other technologies may be a partial explanation – particularly as there are new technical developments and applications announced almost weekly.

But, in fact, my job as Editor has been comparatively easy. First, because the EPIC team – and a few of our friends – pooled their knowledge and practical experiences so willingly. Second – and by far the most important reason – is the dedication and skill of Signe Hoffos who researched, wrote, re-wrote, cajolled, bullied, wrote, researched, interviewed, re-wrote, complained, implored, praised, criticised, wrote, researched and re-wrote every draft.

This book has been a genuine team effort and although Signe has been the star performer I would like to acknowledge the very real contributions of the EPIC interactive technology team, particularly Peter Bazalgette, and Joel Cayford, who read and contributed to the entire manuscript, and also to Gretchen Dankwardt, India Hart, Jan Holmes, Paul Ingham, Susan MacDermott, Jeremy Redhouse, Neera Sarna, and Janet Wood, for their individual contributions. I would also like to thank Graham Beech, Rockley Miller and Michael Segal for their support and assistance.

The final responsibility for producing the book lies with two people. Graham Beech our publisher suggested it in the first place. I agreed, believing that the problems I had encountered in understanding the new technology would have been eased by a book like this. I hope when you've "interacted' with our book you'll agree and share our enthusiasm for interactive video.

Please don't let it stop there. We would like to hear your reactions, plans and experiences. I am sure we are involved in the creation of a new industry and at this stage everyone will benefit from sharing ideas.

Eric Parsloe
January 1984

TABLE OF CONTENTS

ABOUT THIS BOOK

This book is one of the first on a topic which is new and unfamiliar even to people who work in the industries which first made interactive video possible. It will appeal to an audience with a wide range of interests and expertise. For that reason, it has been written with both the domestic and the professional market in mind, for first-time users as well as people familiar with new technology.

On one hand, we are addressing interested amateurs – people with domestic-standard video equipment and personal computers, who want to make programmes themselves, at home, for fun. But there are also business executives with whole networks of industrial-standard equipment and in-house production units – people who want programmes to meet specific needs, and who are likely to commission the actual production work from others. Between these two groups are many others, working with projects, budgets and facilities of all kinds. They all need to know the same things: what can it do? how is it done? what will it cost?

Any study of interactive video can first be broken down into a discussion of the medium's two component technologies: video and computers. It is likely that many of the people who read this book will know a lot about one, and not much about the other. A budding film director and an embryonic systems analyst may seem unlikely bedfellows – but interactive video has thrown them together, and this book seeks to bring harmony to their relationship.

There are many, many books, some of them quite substantial, to tell you about video, and many more about computing.

With the recent rapid growth of home markets in both fields, there has been a blossoming in books and magazines aimed at the new home user – and home users are reaching a high level of competence in both fields. We don't doubt that, given the way interactive video is developing, people who have made programmes at home on domestic-standard equipment will soon have something to teach the professionals.

This book aims only to provide a general introduction to the ideas behind video and computer technology: if you are seriously interested in the nitty-gritty of the hardware, you are advised to go to your local library, bookshop and magazine seller to find more detailed and more up-to-date, technical information than we mean to provide here.

Of course, you don't need to know a great deal about new technology to get into interactive video: with a grounding in the concepts and vocabulary, you can learn much of what you need to know as you go along. (Remember, the technology is still developing, and what is new to you may be new to the old hands, too.)

However, even if you are only planning to commission a programme from an outside production company, it is valuable to have that grounding in the principles and language of the technology. It will help you to follow what's going on around you, and to participate in the project for which you are paying. And, of course, a little spurious learning often comes in handy, especially in dealing with pushy sales people and bores at parties.

HOW TO USE THIS BOOK

There was once a book of nonsense verse for children which had features like the instruction 'Open Other Side' printed on its spine. We hope this introduction is not quite that redundant, but we do think it worthwhile to point out that this book has been written for a wide audience, and is meant to be used flexibly. It is, after all, appropriate to use a book about interactive technology in an 'interactive' way.

As you will see from the table of contents, the book is divided into sections, and the sections into chapters. There are quite a few of these. Each section addresses one aspect of interactive video technology, with the information broken down into chapters the names of which, we hope, are self-explanatory. This should make particular pieces of information easier to retrieve as you work your way back and forth over the material.

The first section is an introduction to the concept of interactive video, and to the words and ideas which you will encounter throughout the rest of this book (and, indeed, in almost anything else you read about the subject).

The second section looks at 'applications'–the ways in which interactive video technology is already being applied. Other examples are scattered throughout the book, to illustrate specific points, but we reckoned that many people would prefer to read about how the technology is being used before delving into how it works. These chapters are not intended as a comprehensive report on the state of the art, but only to give you some idea of what has already been done and what, therefore, is possible.

The third section is devoted to video technology, and the fourth to computers. These outline some basic ideas, and explain key words and phrases in each field. As we've said, these chapters are not as complete, especially in the depth of technical detail, as those you might find in a book dedicated to either subject. They aim simply to explain in general terms how things work, and why technical questions sometimes pose possibilities and problems that don't always occur to people not familiar with the technology. They also point out some of the special demands put on both video and computers when the two are linked.

The fifth section describes the actual equipment that goes into an interactive video system. This is, by the very virtue of being published in a book, a set of guidelines and examples rather than an up-to-the-minute analysis of the hardware market: you will have to go to the trade magazines for that kind of ephemeral information.

The next two sections take you through the planning and production of an interactive video programme. Just by glancing at the chapter titles, you will begin to appreciate how much planning goes into an interactive video programme. Shooting a film is one thing, making an interactive video programme is something else again. This is a guide to making the best use of your resources – and we emphasise that adequate preparation is most important of all.

Finally, we address, briefly, two subjects which are in their different ways equally elusive: How Much Does It Cost? and, Where is It Going?

The section on budgets might be subtitled, How Long is a Piece of String? We appreciate that interactive video projects can be done on a shoestring, or with no expense spared. What we can offer is advice on preparing a realistic budget – and sticking to it.

We approached the chapter on the future of interactive video technology mostly for fun, as an indulgence in speculation after all the hard detail.

You may want to read the chapters in a different order, and to review some and not others as you go along. That, is, after all, one of the ideas behind interactivity – and, as we said, the reason why the chapters are as succinct as possible.

And, too, some overlap is inevitable. It's difficult to know what to introduce first when so many ideas are inter- related. So we've also provided an index and an extensive glossary, and we urge you to use both when you run across a word or an idea about which you are unsure. All key words and ideas are fully explained in the text, but if you do not read the chapters consecutively you may come across something for which you haven't yet read the fullest explanation - so please use the index and glossary freely.

We do recommend that everyone reads the first section: it sets you up for the information to come. After that, you may want to move more freely through the book. So, in the true spirit of interactivity, we have designed a flowchart to guide you. Refer to it as often as you like – that, too, is interactivity.

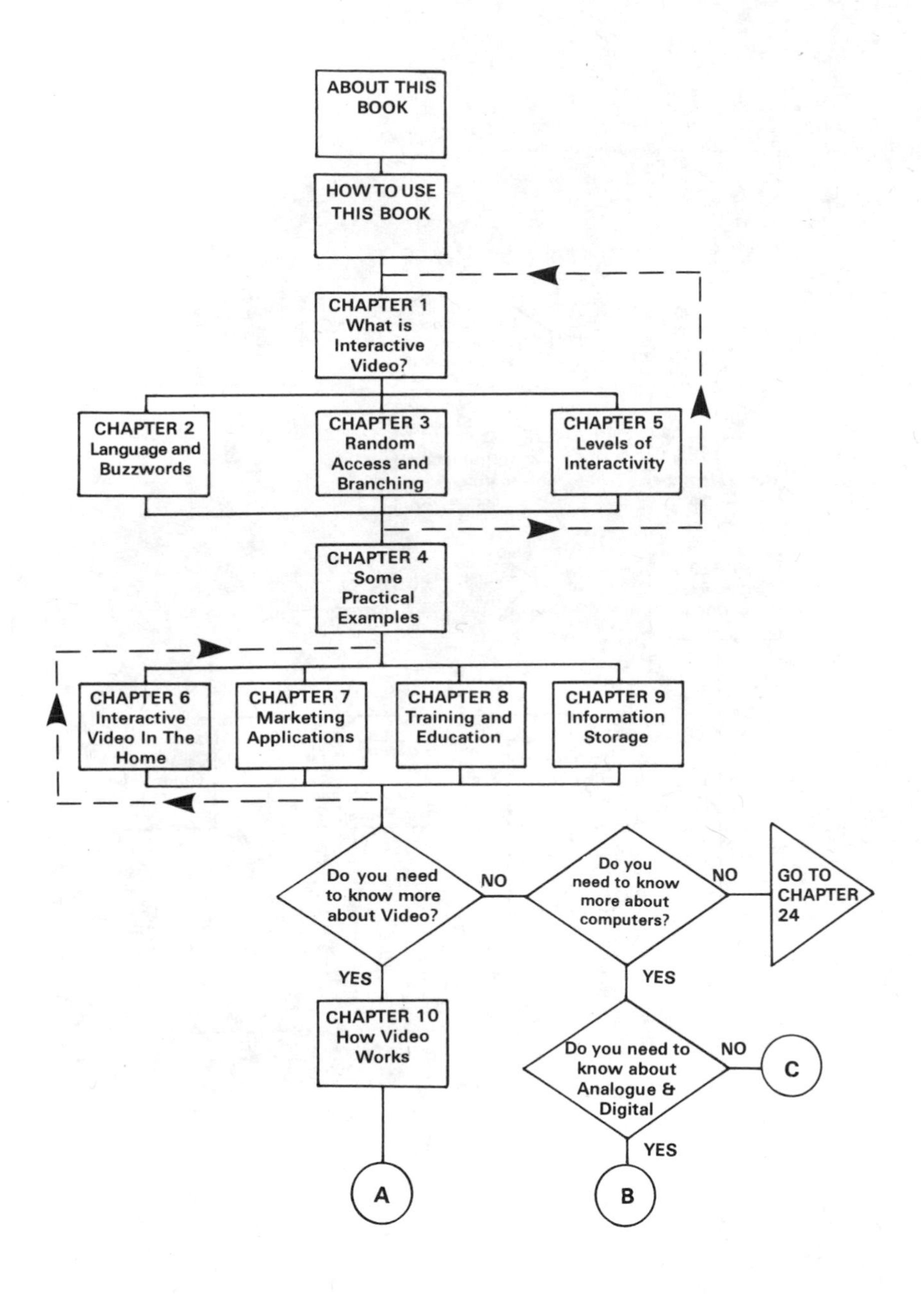

ABOUT THIS BOOK
HOW TO USE THIS BOOK
CHAPTER 1 What is Interactive Video?
CHAPTER 2 Language and Buzzwords
CHAPTER 3 Random Access and Branching
CHAPTER 5 Levels of Interactivity
CHAPTER 4 Some Practical Examples
CHAPTER 6 Interactive Video In The Home
CHAPTER 7 Marketing Applications
CHAPTER 8 Training and Education
CHAPTER 9 Information Storage
Do you need to know more about Video?
NO
Do you need to know more about computers?
NO
GO TO CHAPTER 24
YES
CHAPTER 10 How Video Works
YES
Do you need to know about Analogue & Digital
NO
C
YES
A
B

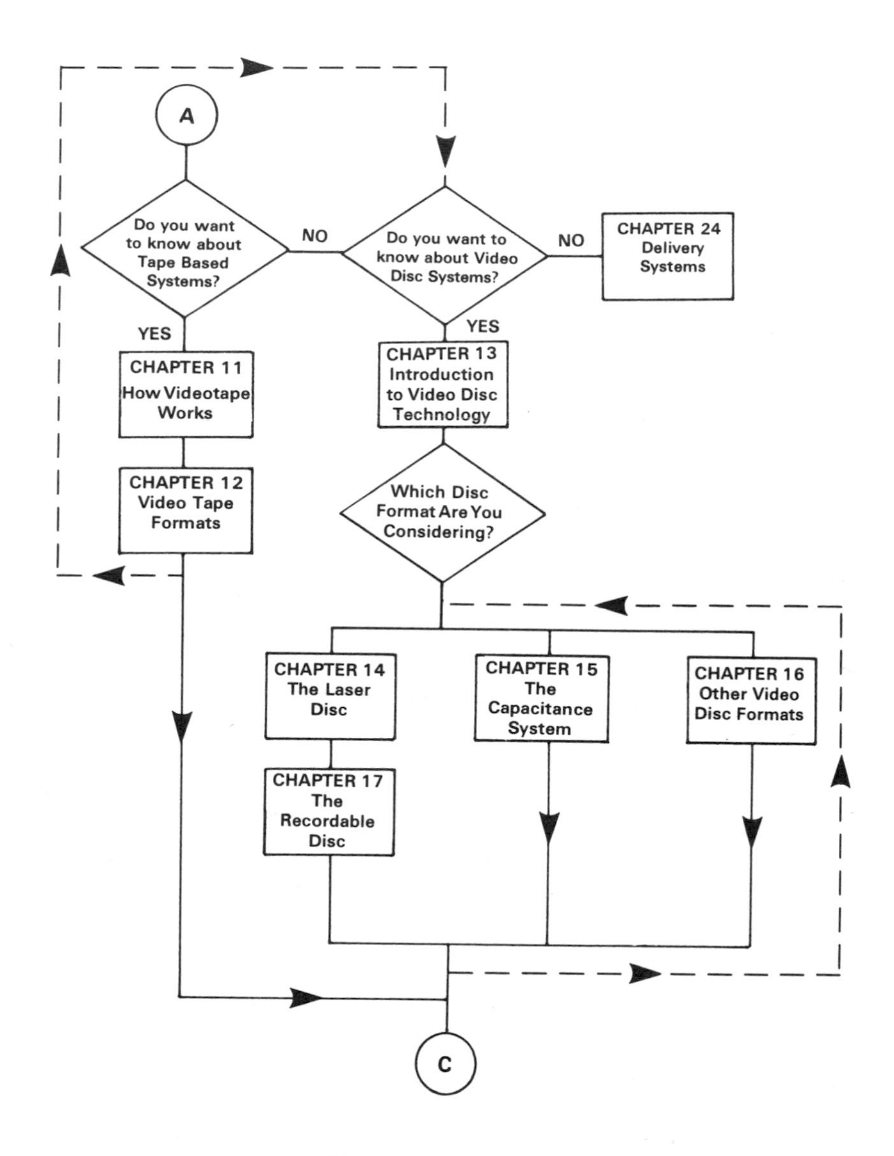
A
Do you want to know about Tape Based Systems?
NO
Do you want to know about Video Disc Systems?
NO
CHAPTER 24 Delivery Systems
YES
CHAPTER 11 How Videotape Works
CHAPTER 12 Video Tape Formats
YES
CHAPTER 13 Introduction to Video Disc Technology
Which Disc Format Are You Considering?
CHAPTER 14 The Laser Disc
CHAPTER 15 The Capacitance System
CHAPTER 16 Other Video Disc Formats
CHAPTER 17 The Recordable Disc
C

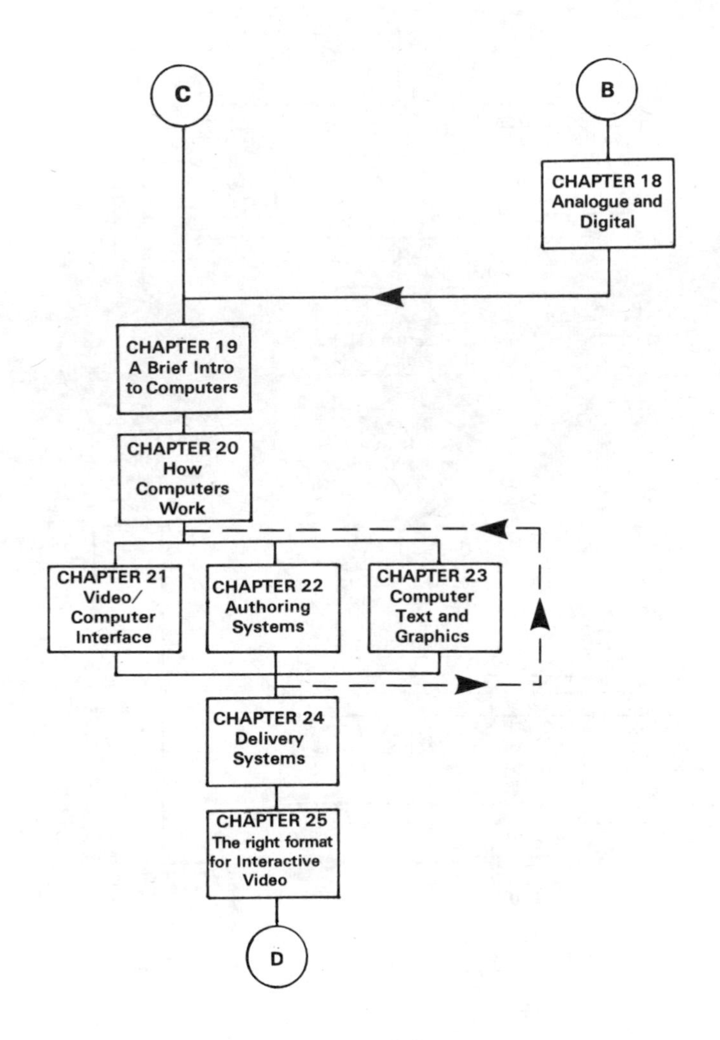
C
B
CHAPTER 18
Analogue and
Digital
CHAPTER 19
A Brief Intro
to Computers
CHAPTER 20
How
Computers
Work
CHAPTER 21
Video/
Computer
Interface
CHAPTER 22
Authoring
Systems
CHAPTER 23
Computer
Text and
Graphics
CHAPTER 24
Delivery
Systems
CHAPTER 25
The right format
for Interactive
Video
D

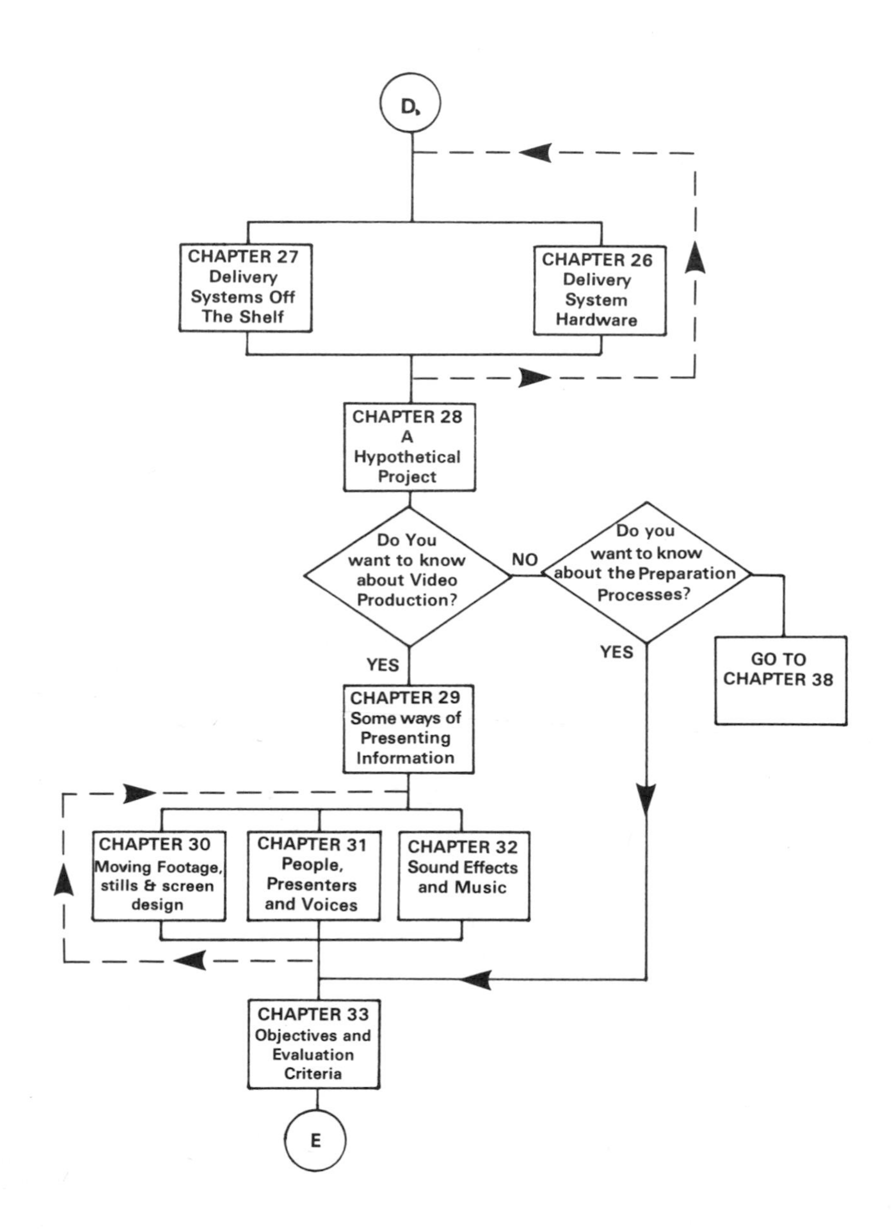
D
CHAPTER 27
Delivery
Systems Off
The Shelf
CHAPTER 26
Delivery
System
Hardware
CHAPTER 28
A
Hypothetical
Project
Do You
want to know
about Video
Production?
NO
Do you
want to know
about the Preparation
Processes?
YES
YES
GO TO
CHAPTER 38
CHAPTER 29
Some ways of
Presenting
Information
CHAPTER 30
Moving Footage,
stills & screen
design
CHAPTER 31
People,
Presenters
and Voices
CHAPTER 32
Sound Effects
and Music
CHAPTER 33
Objectives and
Evaluation
Criteria
E

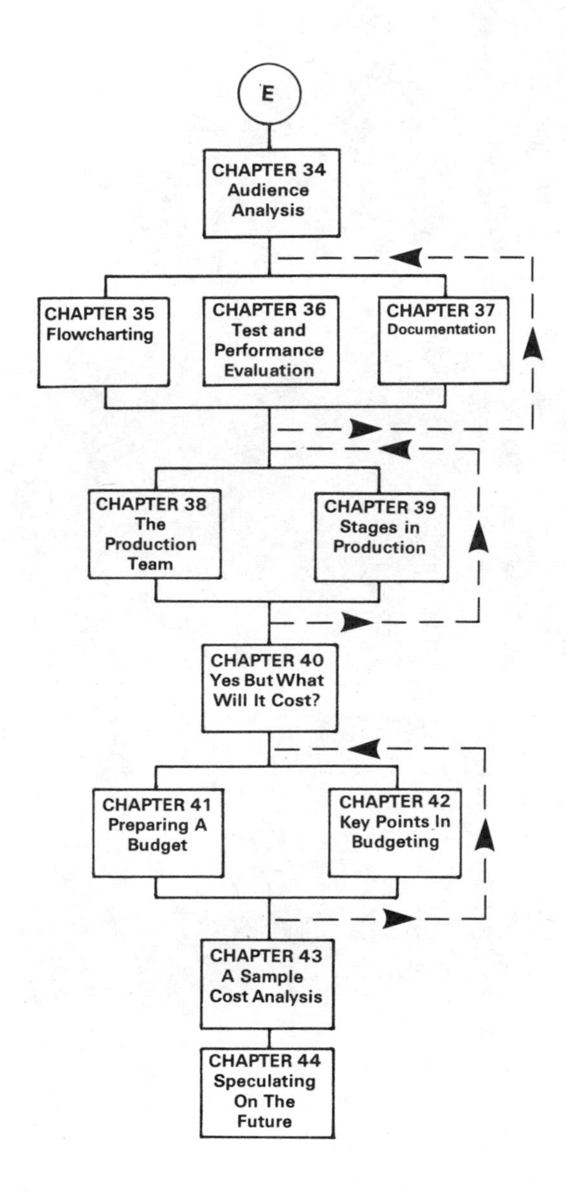
E
CHAPTER 34
Audience
Analysis
CHAPTER 35
Flowcharting
CHAPTER 36
Test and
Performance
Evaluation
CHAPTER 37
Documentation
CHAPTER 38
The
Production
Team
CHAPTER 39
Stages in
Production
CHAPTER 40
Yes But What
Will It Cost?
CHAPTER 41
Preparing A
Budget
CHAPTER 42
Key Points In
Budgeting
CHAPTER 43
A Sample
Cost Analysis
CHAPTER 44
Speculating
On The
Future

CHAPTER 1: WHAT IS INTERACTIVE VIDEO?

What is interactive video? Quite simply, a video programme which can be controlled by the person who is using it. Usually, this means a video programme and a computer program running in tandem. The computer program controls the video programme – and the person in front of the screen controls them both.

The distinction in spelling between the video programme and the computer program is one which we will maintain throughout this book, to emphasise the distinction between interactive video's two complementary, but discrete, technologies.

The idea behind interactive video has been kicking around, at least in academic circles, since the 1950s. From the moment computers first came onto the market, users have been quick to recognise not only their strengths, but also their shortcomings. This desire to enhance the computer's potential lies behind the growth of the computer industry as we know it; it led, over the course of two decades, to the development of interactive video.

Computers can store tremendous amounts of information economically and efficiently. They can handle almost any processing task, completing within hours or even minutes feats of information-handling which would otherwise be monumental if not impracticable. But the kinds of records which computers and people keep have been limited to words, figures and those graphics which the computer itself can generate. While these graphics are now far more subtle and versatile than ever before, in many cases a simple photograph is still far superior to the best of them. Moreover, computers alone are of no use to information which is essentially visual: there are times when a single picture is indeed worth a thousand words.

The tantalising prospect of harnessing computer power to store, sort and display visual images was a driving force behind the development of interactive video. Interactive configurations employing mainframe computers and 35 mm film represent interactivity's dinosaur age. Computer-controlled tape/slide shows - still the right medium in some cases – offer in a limited way what interactive video ultimately achieved.

For it was video, growing up alongside the computer, that proved to be its natural complement. Just as computers entered the top of the corporate market as bulky mainframes, and within a generation had come to the home in the form of user-friendly personals, so video broke out of the technical wizardry of the broadcast television studio to find many new applications first professionally and then in the home. There are now interactive video systems based on the combination of domestic standard video equipment and home computers.

Interactive video represents the fusion of these two pervasive technologies. It harnesses the versatility of computers to the fluency of video, and links two tools with tremendous potential for information storage. Together, a computer program and a video programme can record, store, manipulate and display information from just about any source, from the venerable book to the digital bit.

As a video-based technology, interactive video can capitalise on a wealth of sound and vision – everything, in fact, that conventional film, video and audio recording have to offer: still and moving footage, documentary and archival material, location and studio work, animation, audio recordings of all kinds, graphics of every description. But as a computer-based technology, interactive video has a versatility unequalled in any other audio-visual medium.

In interactive video, the master recording is not the finished programme. It is a storehouse of images and information from which material for any number of new and different programmes can be selected. Any segment, once recorded, can be used again and again, in different combinations which reflect both the overall intentions of the programme's designers and the immediate needs of its users. Material from the video recording can be supplemented by text and graphics generated and stored entirely within the computer.

An interactive video 'delivery system' – the equipment on which the programme finally appears – can employ videodisc or videotape, under the control of either a microprocessor built right into the video player, or a separate computer. The computer program which controls the video programme can actually be encoded on a videodisc, or loaded temporarily into a videodisc player through its remote control keypad, or stored on the magnetic tape or floppy disk of an external computer. In fact, with careful planning, one video programme can be used with several different computer programs to create a number of different interactive video programmes.

But there's more to interactive video than two pieces of disparate hardware talking to one another. In interactive video, the computer program controls the video programme – but, more importantly, the user controls them both. Within seconds of entering the programme, the user who has never touched a computer before can be fully involved in creating the interactive programme appearing on the screen.

Conventional linear video, like television, is a passive medium: the viewer may react to the programme on the screen, but has no more control over it than to turn it on and off. In interactive video, the user controls the actual pace and direction of the programme. It is, in a sense, the user who effects the final edit – every time the programme is run. Far from harbinging a science fiction future where hardware displaces people, interactive video is perhaps the ultimate user-friendly medium: it needs people to give it purpose, or, indeed, to work at all.

An interactive video network might comprise some or all of these elements: a satellite dish, a television aerial, a video camera, a portable videotape player, a videodisc player, a monitor and/or receiver, a computer, remote control devices, a modem and a separate audio system with player and speakers.

On the simplest level, the user, with only a domestic- standard video equipment, can control the speed and direction in which the video programme plays. For example, a home user following, say, a tennis lesson on videodisc, can pause for a slow motion replay when a particular stroke is difficult to follow at speed.

On the other hand, in an industrial application, an engineering worker doing an intricate piece of work can refer to an interactive video manual right on the shopfloor. The apprentice can study the example of an experienced worker performing the same task, and examine the video frame by frame to see exactly how the job ought to be done, then refer to a recorded commentary, or even a set of diagrams, perhaps accompanied by computer-generated graphics showing the latest revisions to the original specification. The unit housing the microcomputer, videodisc player and monitor sits within easy view of the workstations, so anyone on the shopfloor can refer to the electronic manual when extra information or instructions are required.

In the storage of information, interactive videodisc is unrivalled. The avid stamp collector, for example, could put a whole catalogue on videodisc; it would be at least as easy to use as a printed work, and offer much better reproductions. A reference library could store a complete collection, with manuscripts, printed documents, microfiche, photographs and film, safely and compactly on a single videodisc – and retrieve any item within seconds. Reproductions of everything from mug shots to fine art can be efficiently stored on interactive videodisc.

The potential of interactive video is limited only by the imagination of the people who use it. The basic equipment is constantly being refined and improved, but that which now exists is equal to any demand programme designers can make of it. As a teaching tool, interactivity can handle applications on any scale, from an intimate tutorial to a fullscale simulation as vivid and unpredictable as real-life drama. It caters equally for the needs of individuals and small groups. One well-planned programme can adapt to the perceived abilities and needs of many different users.

As the people who design, produce and use interactive programmes, we have most of all to beware of being led by the technology. It is tempting to make programmes that use all the 'whistles and bells' available. The challenge is to approach the problem first, and then to apply existing technology, in a practical manner.

Certain particularly fertile fields of application – marketing, training, information storage – were exploited almost as soon as the technology became commercially viable. The retail applications may be the first to bring the technology into many people's lives, as more and more major retailers introduce interactive point-of-sale units. But many people, too, will use interactive video in training and education, in factories and offices as well as at schools and colleges. Within a few years, interactive video will be as pervasive as the microchip: everything from arcade games to teach-yourself courses will be built around it.

From the relative simplicity of a point-of-sale unit up to the most ambitious training scheme, interactive video brings the same combination of features to every application. These may be found variously in other media, but only interactive video offers them all in one package.

- The same information is always presented in the same way and with the same impact, complete and cogent.

- The user is involved with the programme from the very first – choosing what to view and, at higher levels, interacting with the programme as it unfolds.

* The programme can employ a combination of effects and participants not available to any other type of presentation.

* The user can master basic information through the interactive programme, getting acquainted with new words and ideas, before engaging the valuable time and attention of someone with specialised knowledge.

* A well-made programme can be distributed across the country, or throughout the world.

* With planning, a good video programme can be run with several different computer programs to address different audiences and different purposes.

Interactive videodisc's potential for straightforward information storage is tremendous in itself. The laser disc as we now know it can hold 54,000 frames on each side—enough, the Public Archives of Canada reckoned[1], to record 40,000 pages of text, 5,000 photographs, twelve minutes of moving footage, eight hours' worth of narrated film strip presented at two frames a minute, and 1,000 microcomputer programs, on one digitally-encoded disc.

Video can handle just about any kind of recordable information: print, manuscripts, photographs, slides, film, X-rays, tape recordings, graphics of all kinds—including those generated by computers. Some discs can record both analogue and digital information—in other words, data sent in bits and bytes straight from the computer, as well as information of a more familiar kind. (The whole question of analogue and digital is addressed in some detail in Chapter 18.) Digitised information includes not only computer data, but conventional video and audio signals as well: the potential for telecommunications is incalculable.

IN CONCLUSION

So, these are the basic ideas again:

* In interactive video, the computer program controls the video programme, and the user controls them both.

* Interactive video can employ videodisc or videotape, using either the disc player's onboard microprocessor, or an external computer for computing power.

* Information can be retrieved from any place within the recording and shown at any point in the final programme.

* The user can control both the speed and direction in which the video programme runs—to use slow motion, for example, or to dictate which segment should be seen next.

Of course, there's more to it than that, but the next few basic ideas are presented in separate, short chapters, which should make them easier to read, and to refer back to as you get further into the book. Nevertheless, you're now well away—welcome!

1 Mole, Joseph, and Langham, Josephine. 'Pilot Study of the Application of Video Disc Technology at the Public Archives of Canada'. Ottawa, 1982. (DSS catalogue no. SA2-139/1982) Page 6.

CHAPTER 2: LANGUAGE AND BUZZWORDS

> *"Vague and insignificant forms of speech, and abuse of language, have so long passed for mysteries of science."*
>
> –John Locke, An Essay Concerning Human Understanding

People who, in the 'sixties, talked about 'real time', 'down time', 'dumps' and 'hash addressing' were probably heavily into the contemporary drug culture. People who use those words today are talking about computers. Neither are they the sleek-haired, smartly turned out young men who epitomised high tech in the days of succeeding in business without really trying. Now, your basic computer enthusiast could be a schoolchild, the owner of a small business, a post-graduate student, a medical technician, or even a self- confessed dilettante whose fascination is not with any particular application, but with the technology itself.

In the same way, people who now talk socially about 'back- space editing', 'colour bursts', 'grey scales' and 'lacing', are probably not professional film-makers, but home video enthusiasts.

The last decade has done its fair share for the language, coining words, giving new meanings for old, making catchphrases and buzzwords out of previously stable currency. As much as we've tried to avoid the truly cringe- worthy, you'll undoubtedly run across quite a few new words here in this book. (There is, of course, an extensive glossary at the back, which we heartily encourage you to use as you go along.)

There is obviously a need for new words to describe objects, ideas and activities which are unprecedented in our experience. (It has been observed that one really wouldn't want to say 'great silver bird' when the word 'aeroplane' does much better.) This need can be met by dusting off obsolete or arcane words and investing them with new meanings, or by forming new compounds from familiar syllables, as well as by coining entirely new words.

The word 'car', known to us in association with a decidedly modern conveyance, is a case in point. The word has a long history, coming into English from Celtic through Latin and French, and has in its time described rough carts and gilded carriages. It had somewhat fallen into abeyance as a poetic synonym for 'chariot' before it was appropriated by modern usage as a practical substitute for phrases like 'horseless carriage'. Surely the word will endure long after the automobile as we know it has gone the way of the ancient Briton's two-wheeled wagon.

Undeniably many of the words currently in vogue will be lost as jetsam in a few years' time, while others will enjoy long careers. Some words pass through the language without ever gaining legitimacy; others become common currency because they answer a need in contemporary usage.

Professional fields especially are mined with jargon, buzzwords, compounds and acronyms meaningless to the uninitiated. When one of these reaches the outside world, it is likely to be regarded with suspicion, if not contempt. Computing is notorious for the creation of a newspeak often unintelligible even to other members of the profession. American usage particularly epitomises in many minds the true depths into which instant language can sink under the guidance of the technical minds better at doing than explaining new and complex things.

So naturally, when 'interactive video' first entered the marketplace, many people were understandably dubious, both of the phrase and of the hardware behind it. Interactive video is a fast-growing technology, and the years of its greatest growth have also been those notorious for the assaults of the self-styled technocracy against traditional usage in the language. Unfortunately, the two conditions have combined to produce some commentary on the new medium which is of truly dazzling incomprehensibility.

Of course, computer pundits, particularly, are not above playing on the gullibility of lesser mortals. Secure in their expertise, they will deliberately invent the most outrageous jargon just to see if they can get away with it. One of the contributors to this book once tried to sneak OSVI ('Operator Selectable Visual Information') past people who were both friends and colleagues. His tongue was in his cheek at the time, but the tale is a cautionary one.

Interactive video technology is still relatively new, and there remains a great deal to be learned from experience. Not that this has stopped some people from writing authoritative articles in journals and magazines laying down incontrovertible rules about interactive video technology. A good many of these pronouncements are at best premature, or specific to the experience of one group or person only. Many dictates remain to be proved.

The short answer is, don't be put off, either by the sometimes impenetrable jargon, or by the dictatorial tone of some commentators. The basic idea behind interactive video is sound, and the technology is freely accessible to anyone who wants to exploit it. There is a good deal to be learned from people with practical experience, but you, the user, are ultimately the best judge of what to use and what to leave behind. And if you are confused by a word you don't understand: ask. It might be something you need to know, but equally it might be something coined on the spot that no one else would recognise, either.

There is always a need, in any language, for new words to describe new things: there is no need to use obscure language deliberately to confuse the uninitiated. So, don't be intimidated – and remember, there's a glossary in the back.

CHAPTER 3: RANDOM ACCESS AND BRANCHING

'Random access' and 'branching' are two of the basic ideas behind interactivity, and terms which occur again and again in descriptions and discussions of the technology.

THE BOOK METAPHOR

One of interactive video's close relatives is, arguably, the book. In both, it is possible to identify and retrieve information quickly and directly. In both, the user sets the pace, skimming through familiar material, skipping over irrelevant information, and pausing to study and review new or difficult ideas.

In fact, the two use an index in much the same way, and both also often arrange material in chapters to make separate parts of the work easier to find. And, just as the individual pages in a book are numbered, so in interactive video, each individual frame of video material is numbered. Those numbers are used in exactly the same way in both to find specific references or chapters.

The difference lies in the way in which that search is conducted: in the book, by the eye, the hand and the brain; in the interactive video programme, by the computer and the video player. In interactive video, the time it actually takes to get from one part of the programme to another is called 'access time'. This can vary from a matter of seconds to one of minutes. Both the brain and the computer work at very high speeds; their lightning efficiency is held in check by the tools they are obliged to use – the body in one case, the hardware in the other.

RANDOM ACCESS

So, just as you can find a specific piece of information in a book, using the index and the page numbers in the text, in the same way, using frame numbers, you can find a specific frame or scene within an interactive video programme. You could go directly from any single frame within the programme to any other one, regardless of where the two are stored on the recording. This is 'random access'.

Access time is usually measured at its worst – the longest it can possibly take to travel from one part of the programme to another. (Movement between segments is rarely that extreme – in practice, access time is usually less than the 'worst random access' specification of the manufacturer's literature.)

At one extreme, in a laser disc player where there is no physical contact between the moving disc and the fixed head that reads it, and the disc glides smoothly back and forth beneath the reading head, 'worst access time' is a matter of seconds. At the other, in a videotape player where the tape and the reading head touch one another, and where both move, access can take some minutes, as the two disengage and the tape winds from one reel to the other.

Generally, videodisc offers faster and more accurate access than does videotape. This, coupled with disc's considerable storage capacity, means that a great quantity and variety of information can be put onto one disc, and any one item identified and retrieved within seconds. The appeal of interactive videodisc for information storage is obvious.

BRANCHING

The random access facility is good for more than just calling up archival information, for the computer does more than just flip pages. Random access is the essence of the feature which really makes video interactive: branching.

The word itself is aptly chosen: just as a tree spreads from a single, solid trunk into myriad, ever more slender branches, so an interactive programme moves from the trunk of the main programme off onto branches bearing special scenes and segments. This extra material is usually composed of variations on the central theme, prepared to meet the needs of users with special interests or problems.

Branching occurs from a decision point such as a 'menu' with a choice of what to view next, or a test of some kind. From such a decision point, the programme is prepared to branch to any one of a given number of segments, along paths laid down when the programme was designed. It is the user's choice at the decision point which determines, implicitly or explicitly, which of those paths will be taken.

The user can make a straightforward selection, with a pretty clear idea of what will follow next. Asked 'What would you like to see?" and presented with a choice of six segments, the user selects one, and is soon presented with the first lesson in a training course, perhaps, or a brief product demonstration from a point-of-sale unit. However, the user can also make a decision without knowing what will follow next, in answering a test question, for instance, or making a decision in a simulation exercise.

A simple example of this kind of path is the fork between the right and wrong answer on a test question. The user who gives a correct answer can move straight onto the next segment in the lesson; the one who gives a wrong answer can either be taken through the relevant material again, or shown a remedial segment which approaches the material in a different way. The person who got the right answer first time never needs to see the remedial sequence – only the slower student, who needs the extra tuition, will travel along that branch.

As branching gets more sophisticated, the user may be offered a much wider choice than 'yes' or 'no'. There may be a different remedial sequence for every wrong answer – and a pat on the back for the user who gets the answer right. Commonly, one branch leads to another, and they all weave back and forth around the main trunk of the programme, which contains material common to them all.

Then, too, in a simulation exercise or dramatisation, rather than simply answering questions, the user may be asked to make decisions, and to play out the probable consequences of certain actions. Although the paths that develop here can soon be labyrinthine, the principle behind the branching pattern is the same as for elementary questions and answers. The brief examples in the next chapter will give you some idea of how branching works.

CHAPTER 4: SOME PRACTICAL EXAMPLES OF INTERACTIVE VIDEO

The very next section of this book is devoted to interactive video 'applications', as they're called, arranged in four chapters by category—home use, sales and marketing, teaching and training, and information storage. These few examples are intended only to aid your comprehension of the basic principles by providing some simple illustrations of interactive video in action.

Many applications of interactive video technology, such as information storage, point-of-sale units, or the 'active play' discs of the consumer catalogue, really only use random access to enhance conventional video technology. That is, there is no complicated branching, merely quick moves from one self-contained segment to another.

* In information storage, for example, an entire slide library could be stored on videodisc, one slide per frame, and any one image retrieved simply by punching its frame number into the videodisc player's keypad. The disc player, using its own inbuilt microprocessor, then searches for that frame and displays it on the screen.

* A point-of-sale unit usually offers a choice of product demonstrations or sales promotions. The shopper can see any one by pressing the appropriate numbered key on a pad mounted on the display unit—or, even more simply, by touching the appropriate area on a touch-sensitive screen. The system itself knows on which frame each segment begins, and can interpret a request for information about baby clothes, for example, as a command to search for a given frame number, to run the video from that point to another some distance away, and then to return to a screen offering the customer another selection or, perhaps, supplementary information about credit terms or catalogue sales.

* In a self-teach consumer disc, the amateur golfer can go directly to a lesson on grips and stances simply by requesting the appropriate chapter on the disc through the disc player's keypad: again, the player itself will find the exact frame on which the lesson or chapter begins, and play from that point until it finishes or receives another instruction.

Using the random access facility to branch quickly and easily between disparate segments, an interactive programme can communicate logically and creatively with the person who uses it. A computer program, given a choice of routes to follow, can branch to one and bypass the others; it can store a great deal of information, and draw on only a part of that at any one time. An interactive video programme can hold a whole range of segments, some simpler, some richer than those in the main body of the material, and from these produce a presentation tailored exactly to the needs of the person in front of the screen.

Interactive video is supremely responsive. Individual programmes evolve as each user interacts with the material presented. This does of course open up a whole new range of considerations, functional and conceptual, to the video producer. Interactive video can be quite unlike the linear type, logically or aesthetically. Any one user may never see all the material in one programme, and ten different users may see the same programme ten different ways.

Consider some well-tested applications:

* Prospective car buyers can be helped through a few preliminary decisions by a point-of-sale programme which poses a half-dozen basic questions about things like styling, size, and fuel economy, before presenting a short introduction to the car which seems most likely to suit the buyer's expressed requirements and preferences. The customer with a taste for optional extras is adroitly led to the top of the line; the one looking for a modest family car is shunted off to a familiar domestic model.

* The person going through a simulation exercise will be asked to make decisions: in flight training, one which effects an adjustment in one of the plane's engines; in management training, one which affects an employee's work schedule. Either way, the interactive video programme will branch to present the user with a realistic simulation of the most logical foreseeable consequence of that decision. The plane may soar or crash, the employee may be pleased or enraged. Either way, the user has learned from experience, but nothing irreversible has happened in the real world.

* One of the most celebrated applications was that developed for the American Heart Association to teach the life-saving technique CPR (cardiopulmonary resuscitation). It involves two monitors, one linked to a videodisc player, the other to a microcomputer, and a life-sized 'manikin' (the American equivalent of an English 'dummy') wired to respond realistically to students' practice exercises. After receiving tuition from the videodisc, the student applies CPR to the manikin, monitored by the computer. On the screen linked to the videodisc player, the doctor who gave the initial instructions offers advice and coaching. At the same time, the screen linked to the computer presents a running commentary on the effectiveness of the student's performance, and the condition of the hypothetical patient, supplemented by computer-generated information and text.

* A group at the Massachusetts Institute of Technology (MIT) made a 'surrogate travel' videodisc map of Aspen, Colorado, using thousands of feet of cine film and thousands of still photos to reproduce a complete pictorial representation of the town. With two videodiscs and a wall- sized screen, the programme can project full 360° views of any street there. Using the system's controls, the user can travel through the town, as though on foot or in a car, and even enter public buildings, or pause for a brief anthropological study of the local culture.

These are among the more spectacular examples of interactive video in action - among the furthest extensions of branching we have yet seen. These aren't the kinds of projects which can be put together quickly and easily on a modest budget – but they do represent the limits to which the technology has been taken even in its early years – and the horizons toward which we are looking.

CHAPTER 5: LEVELS OF INTERACTIVITY

How active is 'interactive'? One common means of reckoning is through the scale devised by the pioneering Nebraska Videodisc Design/Production Group to describe interactivity in videodisc players. It is so tidy a classification that it has been widely adopted as a term of reference for both disc- and tape-based delivery systems. There are five levels to the scale, although it is the middle three which most often come into discussion.

It is only the idea of the scale which we are discussing here. How various players, tape and disc, domestic and industrial, adapt to interactivity is discussed elsewhere in the book, and is addressed particularly in Section V.

LEVEL 0: DO NOT PASS GO

'Level 0' would make a handsome addition to the vocabulary of scorn and derision: it's certainly as richly evocative as computing's derogatory 'kludge'. Level 0 describes those video players which are of no use interactively–video disc and tape players which neither have the Level 1 features described below, nor can be used effectively with a computer. These machines are fine for home entertainment, but non-starters as far as interactive video is concerned.

LEVEL 1: RAPID RANDOM ACCESS

Level 1 describes the basic features expected of interactive video players even at a domestic level:

* remote control
* random access (search)
* freeze frame
* forward and reverse
* scan
* slow motion
* step frame
* two audio tracks

At Level 1, remote control is usually effected through the player's keypad, similar to the remote control unit of a television. At higher levels of interactivity, control can be through a computer keyboard, a touch-sensitive screen, a light pen, the paddles and joysticks of video games, or virtually any other object wired to communicate sensitively and intelligently between the user and the delivery system.

The search facility makes it possible to move around within the programme quickly and easily. As we saw in the last chapter, the thousands of separate frames of the video tape or disc are individually numbered, just like pages in a book; the system uses those numbers to find specific frames and segments, just as you use page numbers in a book to find specific chapters or references. This is the 'random access' for which interactive videodisc is especially noted.

Freeze frame and still frame both represent a single frame of video material held motionless on the screen. A still frame is one which has been prepared as a single, static image (an illustration, perhaps, or a page of text) while a freeze frame is one frame out of a length of moving footage, held motionless on the screen. Videodisc is superior to tape here, for some videodisc players can hold a single frame indefinitely, while a tape player can only keep the tape still for a few minutes at a time.

Forward and reverse scan and slow motion make it possible to view a piece of video footage at exaggerated speed. Whether run quickly or slowly, back or forth, the moving footage still appears reasonably coherent. Generally, slow motion is used to examine a segment of moving footage in detail, whilst scan is used to skim quickly through a sequence.

Step frame stands between freeze frame and slow motion: it makes it possible to examine a sequence of video material frame by frame, 'paging' from one frame to the next–rather as one leafs through the pages of a book, glancing at some pages and pausing to study others. Moving footage on video travels at a rate of over two dozen frames a second: with step framing, it is possible to examine at leisure a single frame which would have appeared for only a fraction of a second at its normal running speed.

The two audio tracks can be used to provide stereo sound, but, more importantly, they also offer a choice of separate soundtracks. Simply, this can be used to offer commentary in two languages, or for some arrangement such as commentary on one track, and music on the other. More ambitiously, the two tracks can offer commentary at two levels of expertise, or directed at two audiences: one track in technical jargon, the other in lay terms; or, one directed at selling to customers, the other at training sales staff.

LEVEL 2: BRANCHING

The features expected of Level 1 are really those of 'random access'. True interactivity begins at Level 2, with machines which offer the features of a small computer in a video player.

In addition to the basic features described above, Level 2 offers branching, the facility to move away from the main body of the programme off onto loops and tangents which address the needs of users with special needs and interests. ~~(Branching is described in Chapter 3.)~~

Level 2 also provides the facility called 'score-keeping'. This is a useful tool in many more applications than training alone. Score-keeping can record not only any one person's performance on tests and lessons, but the kind of use a programme is receiving overall. It can be used both to give a student or a class a percentage score on test questions, and also to record in some detail the progress of any one person or group through a programme. This kind of record-keeping grades the programme as well as the student, not only in training but in any kind of marketing or public service exercise. Tests and score-keeping are discussed in Chapter 36.

At the present time, Level 2 really describes the so-called industrial standard videodisc players. These are more rugged, versatile and expensive than either the domestic standard disc machines destined for the home market, or tape players of either domestic or industrial standard. They are designed and built for professional use – in marketing, in education and training, in corporate communications.

Both domestic and industrial quality videodisc players have a small microprocessor onboard – that is, built right into the machine. The microprocessor has the features of a small computer, with a small memory - currently, 5K at the most. (Look up 'K' in the Glossary if you're nervous here).

The circuitry within a Level 1 videodisc player is powerful enough to run the 'Active Play' discs of the consumer market, providing random access, slow motion and so on. The microprocessor inside the industrial standard player is more powerful, and has a larger memory. It can run an interactive program actually encoded on the surface of the videodisc, which is loaded from the disc into the memory of the microprocessor when the disc is put in the machine. A Level 2 player can also temporarily hold a simple program entered through the videodisc player's own keypad.

Much of the appeal of Level 2 lies in the simplicity of the delivery system, which comprises simply a videodisc player, its monitor and keypad. The hardware is easy to display attractively, and to move from place to place, and the equipment is simple to operate. Level 2 is ideal for people whose contact with the system is likely to be brief or infrequent: ~~many point-of-sale units~~, and teaching and training projects are designed at Level 2.

FROM LEVEL 2 TO LEVEL 3

Some inventive things can be done with no more computing power than Level 2 provides, and for many applications the economy of such a delivery system is attractive. However, information encoded on the disc – video, audio and computer programs – once recorded is impossible to change. Many discs carry only one computer program even when they hold enough video information to make up several different ones. Loading a program of any complexity through the videodisc player's keypad is a laborious process – and often a thankless one: the program is lost every time the power is cut off, or when another program is loaded on top of it.

Level 2 technology is impressive, and the very thing for many, many applications. But the limitations described above are serious ones for some users. At the cost of Level 2's compact size, a larger but more versatile configuration was designed: Level 3.

LEVEL 3: COMPUTER/VIDEO INTERFACE

Level 3 represents a video player of any kind linked to an external (or outboard) computer–mainframe, mini or, most commonly, micro. This link can be effected through any one of a number of commercially-available interface packages–which are discussed in more detail in Chapter 21.

Interactivity on tape really exists only at this level: lacking the onboard microprocessor, tape players still need an external computer to effect what some disc players can do using only a keypad.

The computer has a life of its own: programs stored on its floppy diskettes can be changed or amended at any time. Several different computer programs can be built around the storehouse of a single video tape or disc. Furthermore, the computer can generate text and graphics which can be used to supplement the material recorded on video.

In practical terms, this means that the shelf-life of the relatively expensive video programme can be greatly extended through the agency of relatively cheap computing power. On the one hand, as we've noted, one good collection of video material can be used with any number of different computer programs. On the other hand, volatile information such as figures and statistics can be stored in the computer and, so, updated quickly and easily at any time. These can then be incorporated into the final interactive programme as pages of computer-generated text or graphics either interpolated between video segments, or laid on top of the video picture.

Projects which use peripherals such as the American Heart Association's manikins, or the flight simulator's mock cockpit, need the power of an independent computer to co-ordinate all the interactive programme's disparate components.

The full scope of computer/video interface is described in more detail in Section IV.

LEVEL 4: WHAT NEXT?

Level 4 is sometimes used to describe a complete work station–video player, computer, screens, printer, control devices and so forth, and the furniture needed to contain these. However, Level 4 is also used to represent a speculative level of technology as yet unattained by existing systems. If Level 0 might be described as 'no go', Level 4 could be characterised as 'no limits'. What is interesting about Level 4 is that it is discussed at all. Interactive video, itself an innovation on established technology, has allowed room for enhancements into its theory as well as its hardware.

In the final section of this book, we indulge in our own Level 4 speculations. (We feel, though, that we have kept our feet on the ground, and our predictions are closer to Level 3 than Cloud 9.)

SOME EXAMPLES

By way of a quick review, let us just consider the types of programmes you might find at the first four levels of the Nebraska Scale.

Level 0 may be nowhere interactively, but it represents a lucrative field commercially—most of the burgeoning home entertainment market. A film or concert recording that you simply sit down and watch from beginning to end (perhaps giving yourself an intermission, by stopping the player), is an example of Level 0.

Level 1 is represented by 'active play' videodiscs from that same home entertainment catalogue. These are discs with features like chapter stops, still frame collections, two audio tracks, and the facility to run the video backward and forward, quickly, slowly or frame-by-frame. (In some catalogues, these are called 'CAV' discs, a term explained in Chapter 14.) Level 1 discs offer a variety of alternatives to home movies:

* Teach-yourself discs like 'Feeling Fit', 'Play Golf', 'The Master Cooking Course', 'Car Maintenance' or 'Belly Dancing'.

* Special interest discs such as 'Great Railways' or 'The BBC Videobook of British Garden Birds'.

* Performance discs like 'Liza', 'Abba' or 'The New York City Ballet'.

* Entertainment anthologies such as 'The First National Kidisc', 'The History Disquiz' or 'Fun and Games'.

Level 2 is most visible in commercial applications such as point-of-sale units, and in consumer education programmes. Many of the training and teaching discs made both for commercial distribution and for corporate use within large organisations are Level 2. The transfer of a specific quantity of clear, objective information in a relatively straightforward way is ideally suited to Level 2.

The education services sponsored by various drug companies (such as Smith Kline & French's 'Diagnostic Challenges' project) are typical of Level 2 applications. SK&F's system is used in an exhibition travelling to medical centres and teaching hospitals, that can also be taken by the sales force to doctors' offices. A representative programme would be a case simulation, in which the user (in this case, a doctor or medical student) takes the role of the GP following a typical case from the patient's first visit on through examination, tests, treatments and so forth. Using only the keypad of the industrial standard videodisc player, the user can become involved in quite complex programming.

Level 3, as we have already seen, is used in full-scale simulation exercises (such as the American Heart Association's life-saving course) where a computer is needed to co-ordinate responses from a system's several disparate parts. It is also used, more modestly, in teaching and training programmes which employ computer-generated text and graphics to complement or supplement the video programme.

And how is Level 4 used? As we've said, the idea is largely speculative. The best we can say is: wait and see!

CHAPTER 6: INTERACTIVE VIDEO IN THE HOME

One easy way to get into interactive video is through the 'active play' or 'CAV' programmes of the videodisc consumer catalogues.

Active play and CAV are one and the same; the difference between them and CLV, or 'long play', discs is explained in Chapter 14. Basically, laser disc systems currently offer a choice between extra playing time (CLV discs) and interactivity (CAV discs). The rival capacitance systems, VHD and CED, offer a full hour's playing time per side as well as qualified interactivity in one format (see Section III for further description and comparison.)

Since films usually need the extra time, this means that, at least for the time being, you are not likely to find discs that let you re-direct your favourite movies – although you may find CAV discs of films with spectacular special effects. But programmes of other kinds have been designed to allow slow motion, quick scan, step frame, freeze frame, movement back and forth through the disc, a choice of audio tracks and random access to individual segments – that is, 'Level 1' on the Nebraska scale, described in Chapter 5.

The active play discs are still relatively few in number. There is, after all, a great deal more to producing a fully interactive game or entertainment programme than there is to transferring a 35 mm print from cine film to disc. However, theirs is a market with a healthy future. (For one thing, while contemporary economic forecasts are pretty much in agreement that long-term unemployment is here to stay, they do predict a boom in the leisure industries!)

Interactive video, even at this, its simplest level, has a great deal to offer the home user: light entertainment a cut above the standard set by broadcast television, and programmes addressing special interests from art history to karate. Furthermore, it is in the 'active play' mode that videodisc really challenges its rivals on the domestic market – at the present time, neither tape nor television can offer what interactive disc does at this level.

The active play discs released so far fall into four broad categories: straightforward entertainment, sports and physical activites, hobbies and 'teach yourself' projects, and games. Let us look at examples in each group.

ENTERTAINMENT

Although programmes have been made with a view to exploiting the sheer entertainment value of interactive video, a good many entertainment discs are simply souped-up versions of programmes which can be enjoyed in the linear mode, on tape or disc. These are usually music shows or films, and the interactive element really means only that you can jump straight to your favourite numbers, or play action sequences in slow motion (and in silence, if the soundtrack cuts out while the picture is moving at exaggerated speeds).

Of course, disc in any mode does offer excellent sound and picture quality, and resistance to the sort of wear tape suffers in repeated playings–and interactivity, even at its lowest level, adds a dimension not found in any other medium. But when a programme is specifically designed to exploit interactivity - with still frame sequences, slow motion sequences, action sequences designed to be run at exaggerated speeds, creative use of two sound tracks, and so on, it becomes a diversion new to the history of popular entertainment. Some discs do that.

Typical entertainment programmes (principally from Britain, America and Japan) include:

* Popular entertainers, both solo artists and bands, in concert performances such as 'Liza' or 'Pink Floyd Live at Pompeii', or in specially designed collections such as the Abba disc, or Olivia Newton John's 'Physical'.

* Cultural pursuits, such as a typical evening's programme from The New York City Ballet.

* Programmes specifically for children, such as 'The First National Kidisc', a free-wheeling collection of games and activities, 'Fun & Games', or 'BBC Children's Favourites'.

* Light entertainment, such as the 1981 Grammy Award winner 'Elephant Parts', a collection of songs and sketches produced by Michael Nesmith.

* From Japan, light entertainment of another sort: 'Oriental Dreams' ('a sensual portrayal of five young women photographed with great sensitivity') and its sequel, 'Caledonian Dreams' ('the intimate escapades of three beautiful women, two Japanese and one French, as they explore the exotic life-style of the South Seas'), or a disc described in Pioneer's American 1983 Winter Catalogue (and we quote) as 'one of Japan's most famous and adored actresses...presented as the female drama by world renown photographer Kishin Shinoyama in this series of nude image photographs accompanied by a melody of the baroque.'

  (Ironically, when Pioneer took over DiscoVision Associates' California disc-pressing plant, it was suggested that the American site might be used to press blue movies objectionable to the Japanese factory. In fact, the only hint of colour in the US catalogue are these rather dubious Japanese imports.)

* So far unique to Japan, 'karaoke' (literally, 'empty orchestra') discs - recordings of popular entertainers which allow the user to turn off the voice track and sing in place of the star. Very popular in snack bars. (No, we're not kidding.[1])

SPORTS AND PHYSICAL ACTIVITIES

You will learn in Chapter 10 that the first modern application of videodisc technology was the magnetic disc used to provide instant 'slo mo' replay in television sports broadcasting. The facility to review live action in slow motion and even frame-by-frame, which makes interactive video such a valuable tool for manual skills training, also makes it a useful and entertaining medium for sports fans and physical fitness enthusiasts.

The choice of two audio tracks is also a feature here, for it means a segment can be used either with commentary or instructions, or simply with appropriate background music. In a dance or exercise programme, for example, the user can listen to the instructor until confident enough to work to music and pictures alone.

This market divides neatly between the spectator and the participant, with:

* Sports anthologies, many devoted to the humorous side of American football, and to great moments from recent sports history.

* Programmes offering tuition in sports such as tennis, golf, squash and riding.

* Exercise discs such as 'Feeling Fit' and 'Jazzercise'.

* Discs introducing disciplines such as martial arts.

* An intriguing precursor of exercise machines to come, Laser Tour, a bicycle fixed in front of a large screen TV and a laser disc player which projects a biker's eye view of scenic trails as the user peddles along: the faster you peddle, the faster the scenery streaks by. For a change, the bike paths of southern California can be replaced by a roller coaster ride. Laser Tour was launched by the prestigious American department store Neiman Marcus for Christmas 1982; price: $20,000.

HOBBIES AND TEACH YOURSELF DISCS

This could be a burgeoning market if, as many commentators predict, work and leisure patterns change in response to a new economic order, and people devote more time to hobbies and cottage industries.

'Teach yourself' programmes are the commercial extension of the corporate and academic market, but there is every reason to believe that the two will overlap, with adult education institutes, training colleges, the Open University and University of the Air and so on offering interactive disc and tape as well as broadcast television programmes.

The range of these 'special interest' discs already covers:

* Hobbies and domestic pursuits such as cooking, fishing, gardening, car maintenance, photography and 'Training Dogs the Woodhouse Way'.

* Hobbies for observers and enthusiasts, such as 'The BBC Videobook of British Garden Birds', 'Col Culpeper's Flying Circus' or 'Great Railways'.

* Discs devoted to art, and particularly to painting – both catalogues and tours of great collections, such as the National Gallery in Washington, DC, or 'Historic Treasures of the British Crown', and introductions to art history, like as 'The Impressionists', 'Van Gogh' and 'The Adoration' (which compares paintings on that theme by various European masters).

GAMES

All the applications described above, although they lead in turn to fields like marketing, education and training, really draw on only the most basic features of interactivity – easy access to specific scenes, and review at exaggerated speeds (particularly, forward slow motion, step framing and freeze framing); few involve flowcharting or branching in any creative way. It is with games-playing that this higher level of interactivity is achieved by discs within the range of the consumer catalogue. Furthermore, it is likely to be through games that the videodisc (like the microcomputer before it) gains wide acceptance from home users.

As any enthusiast knows, computer games are utterly addictive even with the simplest of graphics (or, in the case of some 'adventure' style narrative games, no graphics at all). The addition of video footage adds to both the effect of the game and the player's involvement with it.

One of the first arcade and casino games was 'Quarterhorse', launched in America in 1982. In it, a video screen shows footage from actual horseraces, while a computer screen displays odds and results for the punters. Races of various kinds have proved popular in arcade and home games. One of the first discs released with RCA's first interactive videodisc player was a game called 'A Week at the Races'.

The extension of this into the lucrative market opened by the 'Space Invaders' craze brings us to disc-based computer games using animation and footage from (or similar to) popular films – particularly, science fiction and fantasy. These games were introduced through the arcades, but their tremendous popularity points eventually to reasonably-priced domestic versions – and the entry of the videodisc player into the home (perhaps to link to the personal computer that came in on the first wave of computer games).

Consider, for example, some of the games unique to interactive video at home and in the arcades:

* Twentieth-century parlour games like 'The History Disquiz', which employs contemporary footage of moments in modern history in a quiz-show format.

* The Mystery Disc series, with titles like 'Murder, Anyone?' and 'Many Roads Lead to Murder'. Classic whodunnits, set in the 1930s, they are replete with the stock characters of the genre: ruthless millionaires and their scheming relations, actresses and society types, playboys and gamblers and, of course, butlers. Players follow sleuth Stew Cavanaugh ('born in the New York slums, worked his way through Harvard') and his assistant Maxie Blair (a 'former editor of Vanity Fair') through the investigation of a crime that has sixteen different solutions.

* Games of science fiction and fantasy, like the mock medieval romp 'Dragon's Lair', intergalactic hijinks such as 'Astron Belt' and 'Star Rider', or the air battles of 'M.A.C.H. 3', that revived the declining fortunes of the arcades games industry.

COMPATIBILITY

With games of this complexity, we come to the one proviso: not all discs are compatible with all disc players – even within the laser disc format. Each model of player carries its own built-in microprocessor. Wholly interactive programmes, such as games, are usually designed to work only on certain players, which will be specified on the sleeve of the disc. (To make the best use of active play discs, players need features like remote control, random access, variable slow motion, and chapter and frame display.) Some programmes are given wide release, in a variety of formats, while others are of limited availability. The short answer is, check the label of the disc first.

DO IT YOURSELF

The next step on from the programmes described above, are the ones you make yourself. Many professionals use 'home made' programmes, based on consumer tapes or discs, to demonstrate interactive video at conferences and seminars. (For many reasons, this type of material is often better suited to a general audience than more technical work.) Discs like the 'The BBC Videobook of British Garden Birds', 'The Romance of the Indian Railways' and even the 'Kidisc' are currently popular sources for this kind of exercise.

The production of such a programme, on tape or disc, follows the pattern laid down in Sections VI to VIII, for the same basic elements are involved whether the end product is an elaborate electronic manual or an amateur's DIY project. The video material could be commercially produced, on tape or disc, but could also be footage you shot yourself.

Such a programme could be:

* The ultimate in home movies: highlights from your last holiday in the sun, or a video 'album' of family and friends – parties, outings, anniversaries, reunions.

* A two-way exercise in games-playing: a challenge both to people who enjoy designing games, and to those who like playing them.

* Supplementary education: passing on old family or cultural skills and traditions, perhaps, or helping children learn in new and innovative ways.

* Promotion or fund-raising for clubs and charities.

These are only a few of the reasons why an individual or a group might enjoy making an interactive video programme; we don't discount possibly the most compelling reason of all (and the one behind the development of much of the technology) – intellectual curiosity.

1 'Made in Japan', Videodisc Monitor, December 1983, and Screen Digest, June 1983, p 105.

CHAPTER 7: MARKETING APPLICATIONS

Marketing is the field in which interactive video has an immediate commercial potential, and it is here that many people will first encounter the technology. However, marketing is also a field in which the tricks and tools of the trade are often cleverly hidden or disguised, so people may at first fail to realise that the intriguing new display at the local shopping centre is not simply the latest wheeze in the scramble for their hard-earned pay, but the vanguard of a technology which will soon be pervasive.

The most obvious application is the point-of-sale unit—the eye-catching display that offers you advice and information, and, somewhere between the product demonstration and the good news about instant credit facilities, persuades you to buy. One stage on from that is the consumer advice centre, which often symbiotically combines information with advertising. A logical development in another direction is the multi-media catalogue, which incorporates still and moving footage on video with computer-generated text and graphics.

Linking marketing to other applications, interactive video emerges as a tool for staff and customer training as well as for consumer education, and as a databank which offers sales people a graceful and versatile presentation tool. Defining marketing most broadly, the applications quickly spiral from the relatively routine to the highly ambitious, from simple point-of-sale units to comprehensive holiday booking services, and from basic consumer education to careers counselling and job recruitment.

The appeal of interactive video to retailers is best illustrated by a simple case study: the debut of Europe's first interactive, videodisc-based point-of-sale unit.

WATCH WITH MOTHERCARE

Mothercare is a chain store devoted to the needs of babies, children and pregnant women. There are nearly two hundred branches in the United Kingdom. The shops bear the stamp of a distinctively British marketing style, neatly stocked with a strictly limited range of quality goods, most bearing the shop's own label. Merchandise includes maternity clothes, nursery furniture, toilet requisites, and toys and clothes for children up to ten years of age—everything from talcum powder to speak-and-spell toys.

In the summer of 1982, an interactive point-of-sale unit was introduced into four shops in England and Wales. It was produced by Realmheath, a London-based firm specialising in video for retailers. The pilot was the prototype of the unit the chain hoped to introduce nationally: a box about the size of Superman's phone booth, with a colour TV screen just above eye level, a simple numbered keypad, and photographs of the products being promoted clearly labelled and numbered below. An industrial standard videodisc player was housed inside the unit.

To see a short video presentation on any one of five product groups, shoppers simply pressed a key on the pad: since the system was disc-based, access was almost immediate. The clips showed products in use, and drew attention to key selling points such as safety features.

Each site was 'twinned' with a branch of similar size, turnover and 'catchment profile', and sales in the eight stores were carefully monitored for twenty-four weeks. About 44,000 people used the unit in that period; it was in use for about a third of the day, and use increased toward the end of the trial period, as people became more familiar with it. Sales figures showed that:

* Sales increased by 20.1% in the four test sites.

* Sales of two of the five products actually fell in the control group, but rose in the test sites.

* The increased sales of just one of the five products 'more than paid for the system'.

* A return on investment of more than 60% could be anticipated in the first year, after writing off the full capital cost of the hardware.

The system was launched in eighty branches nationwide at Christmas 1982, with a new disc containing ten segments ranging from 'Maternity Wear' and 'Safety in the Home' to information about credit plans.

Mothercare's range is admittedly a neat one with which to demonstrate the virtues of the interactive point-of-sale unit, but the salient points are the same for any product range:

* Products can be demonstrated realistically, appearing as they would in everyday use, but set off to best advantage – and guaranteed to be working perfectly.

* Customers can get basic information directly from an accurate and articulate source, before engaging the attention of sales staff, and without undue handling of merchandise and displays.

* The presentation can employ special effects and features not available to the live demonstrator – including the use of well-known faces and voices, and of persuasive background settings and music.

* With demonstrations, voice-overs and perhaps separate computer-generated text and graphics, the customer can be shown a product's full range of features, given current prices and availability, and even instructed on how best to use and maintain the product.

* Each short feature can incorporate both still and moving footage, whether video, photographs, artwork, animation or computer text and graphics.

* A single programme can be distributed to sites of all sizes and descriptions over a wide geographic area – and in regional editions, with different accents and incidental details, if that is desirable.

* When the point-of-sale unit is tied into a computer, the whole sales process, from initial inquiry through to ordering, can be handled on one terminal.

Not the least of these is the simple fact that a well-designed system is fun and easy to use. Even after the novelty wears off, the system remains polite, helpful and informative, and more than a little entertaining.

Of course, there is a footnote. When Mothercare's project began, there were no industrial standard videodisc players available Britain, so American equipment was imported. Designing display units with sufficient ventilation for the machines was a problem at first; breakdown, maintenance and repair was a bigger problem as time went on. Moral? Make sure that the equipment you use is equal to the job – and have a practical contingency plan in any case.

SOME OTHER APPLICATIONS

We have seen a few applications in Chapter 4, and others are scattered through the rest of the book. What follows now is illustrative rather than comprehensive, for marketing applications generally, and point-of-sale units especially, are a burgeoning business everywhere.

Interactive video (on laser disc particularly) has been used to promote products as diverse as desktop computers, earth-moving equipment and ironing board covers. 3M use interactive technology to simulate the appearance of road sign materials in sales presentions. In an enterprising convergence of marketing, consumer education and publishing, a company called VCM Systems produced a videodisc gardening magazine, Greenday Video, in a point-of-sale unit for DIY shops and garden centres. It could hardly fail, and it didn't, clocking up fourfold sales increases for products it promoted.

Let us then look at a few of the projects which fall broadly within the province of sales, marketing and persuasion.

Catalogues for Sales

In the summer of 1981, the American chainstore Sears Roebuck transferred their massive printed catalogue onto videodisc. The 230 page tome was formatted for domestic and industrial standard laser disc players, for distribution to customers' homes and to Sears stores across the US. Familiar catalogue pages (amounting to over 5600 frames) were complemented by sequences of moving footage that included a fashion show and demonstration of a radio-controlled sailplane. Customers found the catalogue great fun, even if the index was a bit daunting. The project was of course a pilot, and waits for greater numbers of disc players in the home market to become commercially viable. But even in the trial run Sears found that with a distribution of about 35 million catalogues a year, the cost of replication on laser disc would be much less than the cost of printing.

Two years later, J.P. Stevens began selling their 'bed and bath' lines through computer-based videodisc catalogues. One such unit together with a small display section can show customers a range of 2000 items in fifty square feet of shop space. Using a retail system from a company called Comp-U-Card, customers can even order goods on the terminal. The whole operation is handled through the dedicated computer, so stores showing the lines don't have to keep goods in stock themselves.

One important publication in the advertising world is the nationwide directory of media rates and data. Britain's BRAD and Canada's CARD, for example, are hefty catalogues in fine print that list the ad rates of a wide variety of media - papers, magazines, TV and radio stations. Now, Britain's 'Mediadisc' is offering that same combination of information and advertising on a laserdisc.

The Car Industry

In Chapter 4, we described a programme that helped car buyers make basic decisions about what they really wanted in a new vehicle. That was the Buick Customer Preference Test, one of the first of its kind on the consumer market.

General Motors pioneered interactive videodisc technology, choosing the medium as the foundation of their marketing, training and communications network as early as 1979, when they introduced laser disc players into showrooms across the US and Canada. It was a momentous step that has since been followed by, among others, Fiat, Ford, and American Motors.

In Australia, GM-Holden found that interactive videodisc addressed the 'economies of scale' unique to a small population in a large country. In Japan, Nissan has a Level 3 system – a disc player linked to a microcomputer – in showrooms, plant and offices, for training and corporate communications as well as marketing. BMW has introduced a purpose-built, tape-based interactive network into its sites around the world, initially for service training, but with potential for other applications. In an early example of creative programming, Ford produced an interactive golf game as a teaser to get staff playing with the system before they brought on the training and marketing discs.

The Travel Industry

'Invitation to Hawaii' is an interactive travelogue produced by Sony (with sponsorship from a group including hotels, airlines and The Hawaii Visitor's Centre) to demostrate what interactive disc offers the travel industry. It features information about sightseeing, sports, entertainment, shopping and hotels, in Japanese or English. The Nebraska Videodisc Design/Production Group in the US, and Brighton Polytechnic in England, are among those who have undertaken similar projects for the tourism and travel industry.

In an interesting application of most existing technology to a single package, the American company Checkpoint Videodisc has developed a system which employs a videodisc player, a Betamax videotape player, a microcomputer and two monitors in a project aimed at the lucrative market for conferences and conventions.

Consumer Education

In the US, Columbia Savings and Loan used interactive videodisc in an attempt to familiarise new clients with the ideas behind saving and borrowing. The system is placed in the reception area, so waiting clients can comfortably get a grounding in basic terms and vocabulary before going into an interview. The client is prepared to discuss business, and the company's qualified employees need not spend their time explaining elementary ideas to new customers.

In neat address of commercial advantage to public benefit, the Canadian company Cableshare developed a marketing tool in which advertising subsidises consumer information. The system is computer-based and can use laser disc. In a typical case, a city newspaper sponsors centres in shopping plazas, to which it feeds data (including news and weather) from a central terminal. The booth features a projection screen and several monitors, displaying advertisements and information – anything from bus schedules to news of the day's best buys. A small shop might take out an ad (using computer text and graphics) on half the screen, and so pay for the information on the other half. A bank might use a videodisc in a longer, more stylish promotion. Either way, the shopper requests information – in fact, about half the users are looking for sales – and sees the ad automatically.

Information as Entertainment

Interactive video technology lies behind a new generation of attractions which inform as well as entertain. Visitors to the 1982 World's Fair in Knoxville saw interactive video in action in the American and Canadian pavilions. NASA use videodisc extensively for information storage, and also in their Visitor Centre at Wallops Island, Virginia. Probably the single most impressive public display of interactive technology is at EPCOT, the Experimental Prototype Community of Tomorrow, a partner to Walt Disney World in Florida. In 1983, EPCOT was showing state-of-the-art technology in 83 programmes, at 272 terminals around the site.

On a smaller scale, Enterprise Square, USA, is an economics education centre designed for the Oklahoma Christian College to introduce visitors to economics through a variety of media including some advanced interactive videodisc programming. Information Technology House at Milton Keynes, north of London, uses an interactive disc made by the BBC and the Open University to show what role technology may play in our lives in the future.

Some quite distinguished museums are using interactive video in the same way as the imaginary museum we discuss later in this book. The International Museum of Photography in Rochester, the Chicago Museum of Science and Industry, and New

York's Metropolitan Museum of Art are among those using interactive video for cataloguing and to introduce visitors and sponsors to the museums and their collections.

Sales, Training and Catalogues

One of the more ambitious interactive video projects yet undertaken was commissioned by IBM to promote their Personal Computer in retail sales outlets across Britain and Europe. The project addresses a very wide range of potential users - first-time buyers, small businesses, large organisations - across several economic and cultural barriers. It also offers a dealer training programme, and a video catalogue.

The first disc, made by EPIC Industrial Communications of London in 1983, uses a lively combination of moving footage and still frames, mime and live action. It runs under the IBM PC itself, and the point-of-sale unit has the added attraction of a touch-sensitive screen (made by Cameron Communications of Glasgow). The first disc was issued in seven languages, with room for 'customising' by country headquarters and individual dealers, through the use of computer-generated teletext on the screen, and 'windows' into the computer program.

More and more, companies using interactive video for sales promotion are learning to use it for cataloguing and staff training too. In another European example, CAT (Computer Assisted Televideo) make a monthly videodisc magazine promoting L'Oréal hair products in German salons. One side of the disc addresses customers, and the other shows hair dressers how to use the products–and how to sell them. (And it's not subtle–the second side opens with a warning to prevent its accidentally being shown to clients.)

Recruitment and Careers Counselling

It may seem cynical to view a graduate's career prospects in the same light as the promotion of hair dye or holidays, but persuasion is the key to both. Many of the same considerations, and hardware, go into the design of a system selling cars and one selling the Army.

For example, in America, attracting students to institutes of higher education is a competitive business. College USA is a project undertaken by a company called Info-Disc to bring sales messages from universities and colleges to high school students and guidance counsellors. Similarly, the Educational Technology division of the US Department of Education commissioned a series on the impact of technology on the job market, to help high school graduates choose careers and training courses. It involves two TV shows, and an interactive videodisc which is being made by the New York company Fusion Media.

One of the largest projects yet undertaken is JOIN (Joint Optical Information Network), an interactive disc-based system designed to supplement the US Army's recruiting programme. JOIN will employ 3161 systems, in a concerted effort to cut through the mare's nest of enlistment options, red tape and bad press which impedes recuitment procedures. The JOIN project has the kind of target figures which would do credit to Mothercare–so we come full circle. It's all selling.

CHAPTER 8: TRAINING AND EDUCATION

"The potential applications of interactive video technology in educational and training fields are virtually unlimited . . ."[1] That is the conclusion of the Council for Educational Technology for the United Kingdom in a report sponsored by the UK Department of Industry and Rediffusion Computers. Already its use is as various as the strengths of the technology itself, and covers a wide range of subjects and ways in which people transfer information.

Interactive video technology appeals as a teaching or training tool in many of the same ways, if for different reasons, that it does as a marketing tool:

* The same information is always presented in the same way: complete, correct, cogent.

* The programme can offer a variety of effects and talent not practical in any other kind of presentation, and offer users information and experiences not otherwise available to them.

* The programme can be shown at any time and in any place – in distant branch offices or rural schools, before a whole class or for a single trainee.

* A well-planned programme can enjoy a long shelf life and wide distribution (and, as an investment, justify devoting extra resources to its production).

* The system can handle basic or routine instruction, freeing teachers and trainers for other work.

* The system can monitor each user's performance, to evaluate both the student and the programme. Supervisors can then identify individual problems, and offer tuition and instruction exactly as needed.

* A well-designed programme is endlessly patient, and utterly non-judgemental: ironically, many students feel more comfortable working with hardware than people when they feel their progress is slow and unsteady.

One way to suggest how interactive video is used in teaching and training is to look at a sampling from fields to which the technology is particularly suited, and which illustrate ways in which information can be presented.

INTRODUCING NEW TECHNOLOGY

It is appropriate to start with a project that uses new technology to introduce new technology. IBM's Guided Learning Centres are perhaps the longest and most widely established use of interactive video in training–and a hearty endorsement from the world's largest computer company. The first Centres were opened in America in 1980, and within three years had reached thirty countries. IBM don't see the Centres as a replacement for 'stand up' instruction, but as an innovation in basic skills training.

Students work in individual study carrels, with hardware ranging from a videodisc player, monitor and keypad to a mock-up of a real work station. (Videodisc is favoured, but tape is used in many countries where the English-language software is not suitable.) Lessons are planned in half-days, so users can balance training with other commitments - and apply what they have learned, in stages, at work. This is less disruptive, less expensive and more cost effective than taking people away for intensive training courses. Fees are based on the anticipated length of the course, but users may be given extra sessions if they need them.

Lessons are a combination of video, reading and exercises, with no more than fifteen minutes of any one sustained activity. Users include both IBM personnel and customers, learning a variety of job skills. Students usually work alone (with supervisors at hand) and while they tend to take the same time in the carrels as in the classroom, they use their time more efficiently working at their own pace.

The same principle has been applied more broadly in the Plato Learning Centres run by Control Data in the USA and Britain. The name is an allusion to the pupil of Socrates - interactive video is sometimes seen as a return to the Socratic method of teaching in that it puts students back in direct contact with the tools–if not the philosophers - that teach them.

INDUCTION COURSES AND PROCEDURAL TRAINING

Among the first to use interactive video technology were banks. The money industry is a likely place to find new technology, for it has both the need and the resources to spend large amounts of time and money on training. People who handle money for a living must meet exacting standards of accuracy and efficiency, and their employers are always on the lookout for new and better training methods. Similar standards apply in many jobs–interactive video has as much to offer white and pink collar jobs as blue collar ones.

Interactive video also addresses another problem. Many employers now find that they attract young people whose academic qualifications are lower than those of their predecessors, and for whom recruiting and training programmes must be restructured accordingly.

* The Bank of America has programmes designed to make bank tellers out of school-leavers who lack even basic book-keeping skills; trainees are taken slowly through elementary principles of banking, and thoroughly tested to ensure that they grasp every concept fully.

* The US Army, among other projects, has been applying the psychology of computer games to basic skills training, and to strategic planning – 'war games'.

PSYCHO-MOTOR SKILLS

'Psycho-motor skills' – hand/eye co-ordination – are a field in which interactive video is unrivalled. Moving footage can record action with over two dozen individual frames detailing every second of activity; with no more than slow motion, step frame and freeze frame, it is then possible to examine any part of a moving sequence in perfect detail. This footage is often better than a live demonstration: it can be shot from any angle, and shown larger-than-life – and it is guaranteed to be a textbook example of how the job should be done, no matter how many times it is repeated.

This type of skills training is easily demonstrated at the domestic level in children's programming (in the paper plane folding or knot tying exercises of The First National Kidisc, for instance) or in Do-It-Yourself and sports programmes. In training, its most familiar form is in jobs relating to maintenance and repair.

MAINTENANCE AND REPAIR: ELECTRONIC MANUALS

Maintenance and repair is not simply a job of making things last longer, but of keeping a sharp eye on mechanical plant of all kinds, from factory equipment to jetliners. Trouble-shooting is constant and critical in industries that depend on expensive, complex equipment. These, like financial institutions, have the resources and the need for training that most immediately justifies the move into interactive video technology. Companies as diverse as Boeing, BMW, Fiat, Ford, General Motors, ICL, Marconi, McDonald-Douglas, Northrop Aircraft, Philip Morris, Rolls Royce Aero-Engines - to say nothing of the armed forces – account for many of the largest and most ambitious projects.

In many industries, printed manuals are worse than useless. They are often so large and unwieldy that workers hesitate to use them at all, and rely instead on a combination of experience, imitation and deductive logic to tackle new or unusual jobs. Not surprisingly, many organisations (including that responsible for civil aviation safety) are keenly interested in new approaches to manuals.

One obvious answer is the electronic manual, which employs video under the control of a computer. The computer also generates text and graphics, which may be displayed over the video picture to explain or modify the information recorded there. Typically, such a manual includes live footage of skilled workers in action, technical specifications and a record (in words and pictures) of changes to these, with still frames, text and graphics, and audio commentary. Most of these are delivered on purpose-built units that include features like customised keypads, touch- or voice-sensitive input devices, and mock-ups of actual equipment.

One prototype of this kind is 'Movie Manual', developed by the Massachusetts Institute of Technology (MIT) as an 'intelligent toolbox' for the maintenance and repair of ten-speed bicycles. It includes live action, stills and both computer-generated and digitally-processed graphics, with a touch-sensitive screen and a customised keypad for easy access to features like 'Stop/Go', 'Zoom' and 'Still' on the player, or 'More' and 'Tools' for auxiliary instruction. In its futuristic mode, it works under the control of a small tracking device which the user wears like a wristwatch. There is even a facility for the user to make 'margin notes' on the screen that can be recorded, amended or erased as if pencilled on paper – the ultimate extension of the book metaphor introduced in Chapter 1.

In industry, electronic manuals are often used as 'on-line' training and reference tools. A typical unit is a trolley or workstation right on the shopfloor. In addition to a monitor (or two) and control devices such as a keypad, touch-sensitive screen or joystick, it either houses a video player (disc or tape) and computer, or has access to these at a central location.

A variation of this is VIMAD (Voice Interactive Maintenance Aiding Device), developed by Honeywell and tested by the US military. Maintenance engineers are equipped with a light- weight helmet containing a headset and miniature TV screen, and a belt carrying a TV receiver, two-way radio and batteries. This leaves the hands free and allows the worker to move about unfettered by wiring; the monitor is seen by only one eye. The worker can 'talk' to the system using keyword commands or a remote control handset. A computer, a videodisc player and peripheral hardware including a voice decoder and a low-level TV transmitter are placed centrally. A senior engineer is on hand to offer advice and assistance. As the helmets and belts are simply display devices, they can be used with different programmes to address various jobs on the line.

Second generation work on VIMAD was undertaken by Perceptronics (who appear again in this chapter). That such equipment is already a commercial reality is some indication of how quickly and dramatically interactive video technology is changing some of the ways in which we think and work.

SIMULATORS IN TECHNICAL TRAINING

Simulation is common in off-line training, in fields ranging from engineering to computer programming. This can be done with a computer alone, but it is really with the convergence of video and computers that simulation techniques have become practical in a wide variety of training programmes.

One fertile field is flight training:

* A training programme at Florida's Embry-Riddle Aeronautical University proved so effective that the US Federal Aviation Administration actually waived minimum flight time requirements for its graduates.[1]

* Vought Aircraft developed a simulator which uses interactive video technology to display various landing approaches for selected flight paths. Both military and commercial airlines use simulators to train cadets and to refresh the skills of air crew.

In Britain, the Trainers and Simulators division of Marconi Instruments has produced work ranging from a full scale reproduction of a nuclear power station control room to a 'suitcase simulator' for flight training. Much of their work is classified; 'public' videodisc-based projects include a fully interactive simulation of the interior of the driver's cab of a metropolitan railway train. (This takes us into the field of surrogate travel, discussed later in this chapter.)

Of course, simulation programmes are expensive to produce, and certainly it is the military's interest in applications of this kind which has channelled a lot of money into the development of interactive technology. But the American Heart Association's course in CPR (described in Chapter 4), with its creative use of media and peripherals, is also a simulation programme, and vividly illustrates the happy ends to which this technology may yet be taken.

SOFT SKILLS

The antithesis of the 'hard' skills addressed by electronic manuals are the 'soft' skills – management, decision-making, and the various techniques which make for good relations between people in polite society.

It might at first seem unlikely that high technology has much to offer a study as nebulous as 'people skills', but human behaviour can be recorded for study as profitably as mechanical skills. The way in which people react to television – not only soap operas and thrillers, but even advertisements – is proof enough of how easily our attention and emotions can be engaged by the small screen. Given the facility to interact with such human drama, people will participate utterly realistically and sincerely – which is just what this technology is all about.

* The Bank of America made a disc called 'People Skills' to prepare bank tellers for situations they will face on the job, from the irate customer to the con artist.

* VISTA, the US Army's Videodisc Interpersonal Skills Training and Assessment project, addresses a variety of problems (such as verbal abuse and insubordination) which commonly face non-commissioned officers.

* Britain's Royal Military Police, with a three-man team and a studio they built themselves, use interactive videotape to train new recruits in policing skills.

* Lehigh University's Instructional Technology Centre developed a series of interactive tape programmes to help teachers of learning-disabled children.

* American Express UK together with IVL Learning Systems made a tape-based programme called 'Cash Flow Problem', a dramatisation in which users make a series of management decisions, either freely or against a structured system. At the end of the tape each user's performance is set against a stored programme to assess individual management style.

LIFE SKILLS

There are skills even more basic than those described in personnel management and customer relations: the 'life skills', which present a challenge to the learning-disabled which people without physical or intellectual handicaps can hardly conceive. This, too, is a field in which interactive video addresses a very real need in a new and dynamic way.

* The Nebraska Videodisc Design/Production Group made a disc on problem-solving for hearing-impaired children, presented in the style of a detective game.

* UCLA (the University of California at Los Angeles) developed a system of language training for the hearing-impaired using discs from the consumer catalogue along with one displaying sign-language phrases, under the control of a computer which also generates text and graphics.

* Utah State University's Exceptional Child Centre made a series of discs to teach concepts like 'Matching Colours, Shapes and Sizes' and 'Timetelling' to the mentally retarded.

* John McKeirnan, who teaches mentally-retarded teenagers on the Isle of Man, uses low-budget domestic videotape and computer equipment, live footage, dolls and models which his students make, to simulate life skills (such as a visit to the corner shop) through play.

  Such projects display interactive technology's unique strengths in addressing users whose needs are more critical than those of any engineer or factory worker.

EDUCATION

The applications discussed so far represent projects undertaken by large organisations for their own use or, in the case of the life skills programmes, pilot studies developed in academic environments. It is in the field of education that we find interactive 'courseware', the high tech answer to textbooks. 'Generic programming' is material which can be used as widely as, say, educational television or commercial training films.

Medical applications, and the work of the drug companies in particular, stand between the public and the private. There are on the one hand education projects like the American Heart Association's course in CPR (described in Chapter 4) and 'A Diabetes Primer', developed at the University of Texas by Jim McBride and Rick Kent. There are also projects sponsored by drug companies which, while educational, obviously have a secondary purpose. These often employ a variety of effects, including dramatisations, to simulate diagnostic and clinical procedures. Most are aimed at medical students and qualified personnel. Some, like Miles Laboratories' 'Learning Centres' series, are offered simply as courseware, while others, like Smith Kline & French's 'Diagnostic Challenges' programme (described in Chapter 5), are mounted in travelling exhibitions which can be taken to teaching hospitals, medical centres and doctor's offices.

Other projects for education represent co-operation between government agencies, universities and private enterprise.

* The Nebraska Videodisc Production/Design Group have made many programmes, including a series on laboratory sciences for post-secondary students, a disc for physics students called 'The Puzzle of the Tacoma Narrows Bridge Collapse', and one on whales made with the National Geographic Society. 'Evidence Objection'

is a role-playing exercise for law students which not only helps them recognise inadmissible evidence, but also gives them a feel for courtroom procedure which would be difficult to recreate in any other way.

* The 'Schooldisc' is a project undertaken by the American Broadcasting Company (ABC) and the National Education Association (NEA) to make programmes for primary schools.

* The Minnesota Educational Computing Consortium (MECC) made a course on economics specifically for rural schools where student numbers are too low to justify any but the most basic subjects in the curriculum.

* Cyclops, a system developed by Britain's Open University, has been used variously for distance learning (using telephone lines), language training and teaching children with learning disabilities.

* One of the first discs produced by Thorn EMI for the VHD system was 'Start Here–Adventures in Science', an educational programme (showing experiments that can be performed in the kitchen) that was received well by kids, if with some trepidation by parents.

* One of the first discs made for RCA's CED system was 'A Walk Through the Universe', produced by the CBS Publishing Group to introduce astronomy to students.

* ActionCode is a Level 3 system from Perceptronics, marketed by the US National Education Corporation. The hardware includes a videodisc player, a microcomputer and coded notebooks which are read by a wand, rather like products in a department store. The courseware is aimed at technical training–electronics and so on.

The Council for Education Technology, in the report quoted at the beginning of this chapter, cites many fields in which generic programming could be used. Ultimately, we may see distance learning through teletext, satellite and cable, with video tape and disc as well as printed material used to supplement broadcasts, and students receiving both video and computer programmes on domestic standard equipment. The technology exists–we need the resources to exploit it.

SURROGATE TRAVEL

Finally, we address a rogue category, part training, part information storage, part entertainment. In Chapter 6, we mentioned Laser Tour, a home exercise machine which projects passing scenery in front of a stationary bicycle. This is 'surrogate travel' in its most innoculous form. In Chapter 4, we described MIT's 'movie map' of Aspen, Colorado–a more startling programme. Earlier in this chapter, we cited Marconi's simulator for train drivers, which projects realistic moving footage through the 'window' of the cab.

Surrogate travel allows users, anytime and anywhere, to move easily through places to which–for considerations of distance, time, security, cost or bother–free access is undesirable or infeasible. People from many disciplines are interested in the idea of vicarious travel, whether in a building, a town or through open countryside, entirely through machines. Not surprisingly, the military were quick to pick up on the idea.

Surrogate travel systems typically employ at least one videodisc player under the control of an external computer, with devices such as joysticks or simulated control panels to monitor the speed and direction of travel.

* Perceptronics have developed several programmes for the American armed forces; one helps to train operators of shipboard ammunition elevators, another shows trainees the appearance and layout of places they cannot visit.

* Both Perceptronics and MITRE Corporation have created display systems which allow users to study tactical terrain maps at different levels of scale and detail.

* In Britain, Marconi have also developed a variety of programmes, principally for military use, for surrogate travel indoors and across terrain.

These are only a hint of the ends to which surrogate travel can be taken. (We are, after all, considering only the use of footage which has been obtained legitimately.) This could be the armchair travel of the future – it is certainly a powerful tool for military strategy.

These examples suggest in broad terms what has already been done with interactive video technology – and what could yet be done. How our perceptions will change if we depend principally upon a screen for information about the world about and beyond us, is another question altogether. Some speculation about the future is indulged in Chapter 44; to know what really happens, we will have to wait and see.

1 Duke, John. Interactive Video: Implications for Education and Training. Council for Educational Technology, London, 1983. Page 104.

2 'An Interactive Video Casebook', Educational & Industrial Television, June 1982, p 39.

CHAPTER 9: INFORMATION STORAGE

In interactive video, as in computing, there are two types of application, the active and the passive—or, in the words of David Hon, 'the dynamic and the static'[1]. The most innovative projects, like Hon's course in CPR for the American Heart Association (described in Chapter 4), may be the dynamic, but it is in the static—in information storage—that interactive video, like computing before it, is likely to make some of the most telling changes.

'Please do not bend, spindle, fold or multilate' was a popular catchphrase in the hoary old days of key punching, when much light humour was devoted to a vision of life under the domination of robots and paper tape. Ironically, it was in the ancient guise of games-playing that the computer finally entered the family circle, and the clock radio is about as far as most people have come under the domestic tyranny of high technology.

What has happened is in fact far more subtle than anything predicted by the glossy magazines and saucy cartoons of the 'sixties. Microchip technology has indeed gone into making domestic applicances fancier than ever before but, far more insidiously, it has also admitted tremendous changes in the ways in which we handle information.

As we have seen, while they are great workhorses, computers alone cannot handle information that is essentially visual. However, when computers are linked to machines that have the capacity for storing visual information that computers have for data storage and processing, a powerful new tool is created. The innovation of interactive video (particularly, that involving optical digital technology) answered this need. Vast amounts of data from many sources can be gathered, stored, processed and disseminated through systems employing some aspect of magnetic or optical disc or tape technology. Records which once filled a building now fill a room, and information-handling tasks which might once have appeared monumental, if not impossible, can now be completed with hours or even minutes.

OPTICAL DIGITAL DISC

The development of optical disc technology (of which the videodisc is only a part) has brought to computers a storage medium more versatile and efficient than any that has gone before. These discs are computer-oriented; they can encode not only sound and vision, but information of virtually any kind through the digital storage techniques developed in computing for use with magnetic storage media. (Analogue and digital ways of handling data are discussed in Chapter 18.)

Dozens of companies have undertaken development work on optical disc systems, and this number includes most of the heavyweights in computing, communications and audio/visual technology. Competition is fierce, and security has been tight throughout the development period.

As their purpose is to store information, optical discs have to be at least as stable as videodiscs, so one critical question is the actual material from which the discs are made. Metallic and photographic films, compounds of light- sensitive metal and plastic, and even organic dyes have been considered. Different ways of encoding the disc include refinements of established laser disc technology, but also consider other ways of inscribing patterns. (One potential method is descriptively called 'blistering').

The quest for a perfect 'clean air' environment is another goal in optical disc technology. There is a touching irony in the fact that atmospheric pollution—good old-fashioned dust—is a disc's worst enemy: a single microscopic speck in the equally microscopic recording track will corrupt the information held there. This can mean distortion of the video picture, or corruption of the computer program, which is why all discs are coated or housed in plastic. (Even the recordable discs described in Chapter 17 are coded by lasers which penetrate their protective plastic outer layer.)

HIGH DENSITY OPTICAL DISC STORAGE

We referred in Chapter 1 to a calculation from the Public Archives of Canada, that 40,000 pages of text, 5,000 photographs, eight hours of film strip, twelve minutes of video and 1,000 computer programs would fit onto one digitally-encoded disc.[2] The striking example often quoted to demonstrate the optical digital disc's capacity is the prospect of recording the entire Encyclopaedia Britannica—43,000,000 words and 30,000 illustrations—on a single disc the size of a conventional audio LP.[3]

This represents storage capacity ten to a hundred times greater than that of contemporary magnetic tapes and disks. It also represents a medium which, once initial change-over or start-up costs are covered, should be the cheapest yet. (A chart contrasting the relative costs, at 1982 prices, of the media needed to hold 10^{11} bits of data—the capacity of one 12" optical disc—starts off with US $40,000 for eighty packs of magnetic computer disks, and works down through $1350 for ninety computer tapes, and $100 for 2400 feet of high density magnetic tape to $10 as the projected unit price of an optical disc in a large run.)[4]

In digital transfer, an image such as a page of a book is optically scanned, and the information it contains (textual and graphic) broken down into binary digits—zeros and ones, the pulses which are the basis of digital technology. This information can be stored, processed, transferred to other media, broadcast over great distances, and finally reconstructed in its original form without noticeable deterioration. The digital pulse ('on' or 'off') is a brief one, so data can be packed very tightly onto a digitally-encoded disc (often, through the same pattern of microscopic pits used in familiar laser disc technology).

Interactive technology can employ media other than optical discs, but those originating solely in video or computing (such as videotape and floppy disk) are only a stage in the transition to the kind of comprehensive storage systems optical disc presages.

Magnetic and photographic media are subject to deterioration, and no system so far offers the combination of features that optical disc does. Even an analogue-encoded videodisc offering 54,000 frames per side could hold the equivalent of 1,350 slide carousels of eighty slides apiece on one double-sided disc.

Of course, high-density storage on any medium is of little use without some rapid and responsive way of retrieving information. This is where interactivity comes into play. With an information retrieval system which is both efficient and versatile, any single entry could be found with minutes or seconds, from any one of a variety of approaches.

INTERACTIVE VIDEO IN INFORMATION STORAGE

The Public Archives of Canada's initial report[5] noted some of the salient points in the archival use of information:

> "The physical reality of documentation varies from parchment to paper; from photographs to print; from paintings, drawings, medals and maps to audiotape, videotape and film."
>
> "Direct handling of original materials by users . . . can severely reduce their longevity . . . ."
>
> ". . . on the whole, most users of archival material are not concerned with the physical nature of the media; it is the purely informational content that is of interest to them . . . ."
>
> "The archivist's most pressing problem is to protect the original while making the informational content readily available."

Consider, in response to this, an information storage system employing optical disc under the control of an external microcomputer. It would offer:

* Integration of material from virtually any source medium, moving or still—paper, tape, disc or film - into one compact, stable package.

* Extremely high quality reproduction, especially of visual material taken directly from 35 mm cine or slide film, or good quality videotape, or data generated directly from a computer.

* Protection from deterioration, through wear or aging, of picture quality or the accuracy of signals, from the moment the information is first digitally encoded.

* Phenomenally high density of information storage: optical discs stacked and read as magnetically-encoded computer hard disks are could easily put storage into the range of 'terabytes'. (That is, 1,000,000,000,000 bytes—'tera', meaning 10^{12}, being derived with so irony from the Greek word for 'monster'.)

* Accurate random access, usually within seconds, of information encoded on any part of the disc (accurate to a single frame of visual material).

* A wide choice of information retrieval programs and tools, from hardware and packages designed by commercial distributors to purpose-built systems made to suit specific applications.

An information storage system based on interactive video technology promises to cut across traditional methods of classifying and disseminating information, in archives and libraries of all kinds. Consider what effect the widespread use of compatible systems, based on optical disc and external computer, could have on access to information:

* Reference material need not be divided between books, objects, photographs, film and so forth, all indexed and stored separately, but could be collected according to subject matter, with information from a variety of source media recorded on a single disc.

* With computer-based indexing systems, access to a great variety of information can be given quickly, compactly and creatively, offering users a variety of approaches (alphabetic, numeric, keyword) to any collection.

* Such information programmes can be widely distributed - as a commercial proposition for hard-pressed research and educational facilities. No less an institution than the University of Cambridge is exploring the idea of producing discs for sale to other establishments. The Centre for Aerospace Education at Drew University has made a series of 'Space Discs' with footage from NASA and other research bodies and observatories.

* The perennial problem of storage suddenly becomes much simpler: hard copies of books, documents and so on can be stored safely and cheaply in distant locations, while duplicate information is offered safely and compactly for reference on disc.

A great deal of research time is spent trying to get to materials – searching for sources, learning where material is held, wading through indexes, securing access to private collections, waiting for material to come from distant stores. These problems face research students and maintenance engineers equally – and they are addressed by interactive video technology.

In addition to the Public Archives of Canada and Cambridge University, the British Library and the US Library of Congress, and the French national military archives, are among the institutions investigating the use of interactive video technology in information storage. The emergence of recordable disc technology (described in Chapter 17) is aimed at information storage within institutions of this size, but the price of the hardware will inevitably come down, and DIY disc storage will become widespread.

As a sample of what interactive video can do for information storage, let us look at one early, typical, application.

VIDEO PATSEARCH

Video Patsearch was developed by Pergamon International Information Corporation. It is a Level 3 system designed specifically to facilitate patent searches. Textual information is kept on a database on the Bibliographic Retrieval Services (BRS) search system, and is accessed by telephone lines on a time-sharing system.

On the computer screen, the user sees the abstract usually found on first page of patent document, which can be printed in hard copy. Supplementary graphic information comes from the videodisc and is displayed on a separate screen. The design offers 'single key access' through time-consuming stages of the process, and temporary storage of material to which the user wishes to refer again. Users can work with a number of keywords and dates to find information they want.

When it was released, in 1981, Video Patsearch offered over 720,000 American patents issued since 1971, on eight laser discs (a rate of one frame per patent). The service is designed to expand at the rate of at least 50,000 patents a year, with weekly updates to the database, and a new disc every quarter. In 1981, subscription cost $6000 a year, with pro rata charges for time used on-line (typically, about $10 per search at that time). The subscription price includes a purpose-built unit with two screens, a keyboard, a printer, a Pioneer industrial standard disc player and the Video Patsearch computer terminal.

There have been considerable refinements to the available hardware even since Video Patsearch was released. So, as much as Video Patsearch represents a significant model early of integrated technology, it is also only the first among many interactive video systems which will revolutionise information-handling in the years to come.

1 David Hon, Director of Advanced Technology, American Heart Association, speaking at Computers & Video Convergence 'State of the Art' Conference, London, October 1983.

2 Mole, Joseph, and Langham, Josephine. 'Pilot Study of the Application of Video Disc Technology at the Public Archives of Canada'. Ottawa, 1982. (DSS catalogue no. SA2-139/1982) Page 6.

3 John Free, 'Optical Disc Can Store an Enyclopaedia', Popular Science, August 1982, page 47 ff.

In fact, in 1983 Grolier Electronic Publishing of New York, together with Britain's Longman Video, began work on an electronic encyclopaedia, using computer database and laser disc to provide text, audio and video.

4 Public Archives of Canada, loc. cit.

5 International Resource Development Inc (Norwalk, Connecticut), quoted in Popular Science, op. cit.

CHAPTER 10: HOW VIDEO WORKS

Largely for historical reasons, different, incompatible ways of composing and transmitting a television signal have developed in different parts of the world. These are generally called 'production standards', or, simply, 'standards'. Of these, the most important are the line standard, the field standard and the colour standard.

In practice, the three are effectively bound together. There are now basically two line and field standards, and three colour standards. These are combined to form three basic packages which, largely for political reasons, have been adopted variously around the world – although many countries are still phasing out older, more diverse standards. The three systems are called by the initials which properly describe only their respective colour standards. They are:

* NTSC (established by the American National Television Standards Committee and employed in the USA, Canada, Mexico, the Bahamas, Japan and the Philippines);

* PAL (Phase Alternation Line, developed in West Germany and adopted in the UK, through most of Europe, the Middle East, Africa, Australasia and South America);

  and

* SECAM (Séquential couleur à mémoire, developed in France and used also in the USSR and its satellite states, and in some Middle Eastern and African countries).

The components of the three basic packages will be explained as we go along. Of course, there are variations, but in general NTSC, PAL and SECAM describe the most common configurations. In fact, it's common practice to generalise still further and lump SECAM in with PAL, as they are similar and PAL is by far the more widely used. So different, incompatible systems are often described only in terms of the distinction between PAL and NTSC.

What follows is a brief explanation of the three standards, and of how the picture on the screen is composed and transmitted. You can of course find ever more detailed explanations the further you go into the relevant technical literature.

THE EYE, THE BRAIN AND THE TELEVISION

Whether moving or still, the image which we perceive as a single, integrated picture on the TV screen is in fact like a sleight of hand, an idea agreed between the eye and the brain when both are working with illusory information.

The eye (not unlike the videodisc player) interprets much of the information it receives as variations in light, and it needs about a fifteenth of a second to register an entirely fresh image. When things happen faster than that – the eye and the brain have difficulty following every motion in perfect detail.

This is why, for instance, the wheels of a speeding bicycle or the figure of an athlete in motion often leave us with a blurred impression rather than a sharp image. This is also why sporting artists had little clear idea of how a galloping horse would really look, frozen in mid-air, until photography came along to record every stride in split-second detail. If you compare an old painting of a racing horse with a modern photograph of one, you will see that there is little resemblance between the rocking horse straddle of antique art and the motion of the real animal.

In the world of cine film and video, the result of all this is that we are, happily, tricked into seeing 'moving pictures' where, in fact, there is in fact only a long series of still images moving at high speed.

So, the image on the screen is a grand illusion – the trick works because it is done so fast. In fact, it is truly deceptive, for what begins as something concrete – a scene in a television studio, perhaps – and is received by us as sight and sound, is actually turned into electrical signals, invisible and inaudible to our senses, while it is being recorded and transmitted. To understand what the different standards represent, let us first look at how the video picture is composed and transmitted.

RECORDING AND TRANSMITTING VIDEO

Television 'sees' in much the same way that people do: it assembles a detailed picture in black and white, and then applies colour to it. Colour blindness in people is a different thing from near-or short-sightedness: in eyes and television both, a picture and its colour are complementary but separate pieces of information. And, just as images focused on the retina by the lens of the eye are transmitted to the brain as nerve impulses, so images passing through a camera tube are converted to electrical signals.

The picture on the screen is rather like a photograph in a newspaper, composed of innumerable tiny dots which the eye and brain interpret as one coherent image. In a sharp newspaper reproduction, there are literally thousands of tiny, subtly graduated dots in long, uniform rows which, at a short distance, look plausibly like smooth bands of light and shade. In a grainy newspaper reproduction, these become obvious even to the untrained eye, while in a glossy colour magazine, they are virtually unobservable.

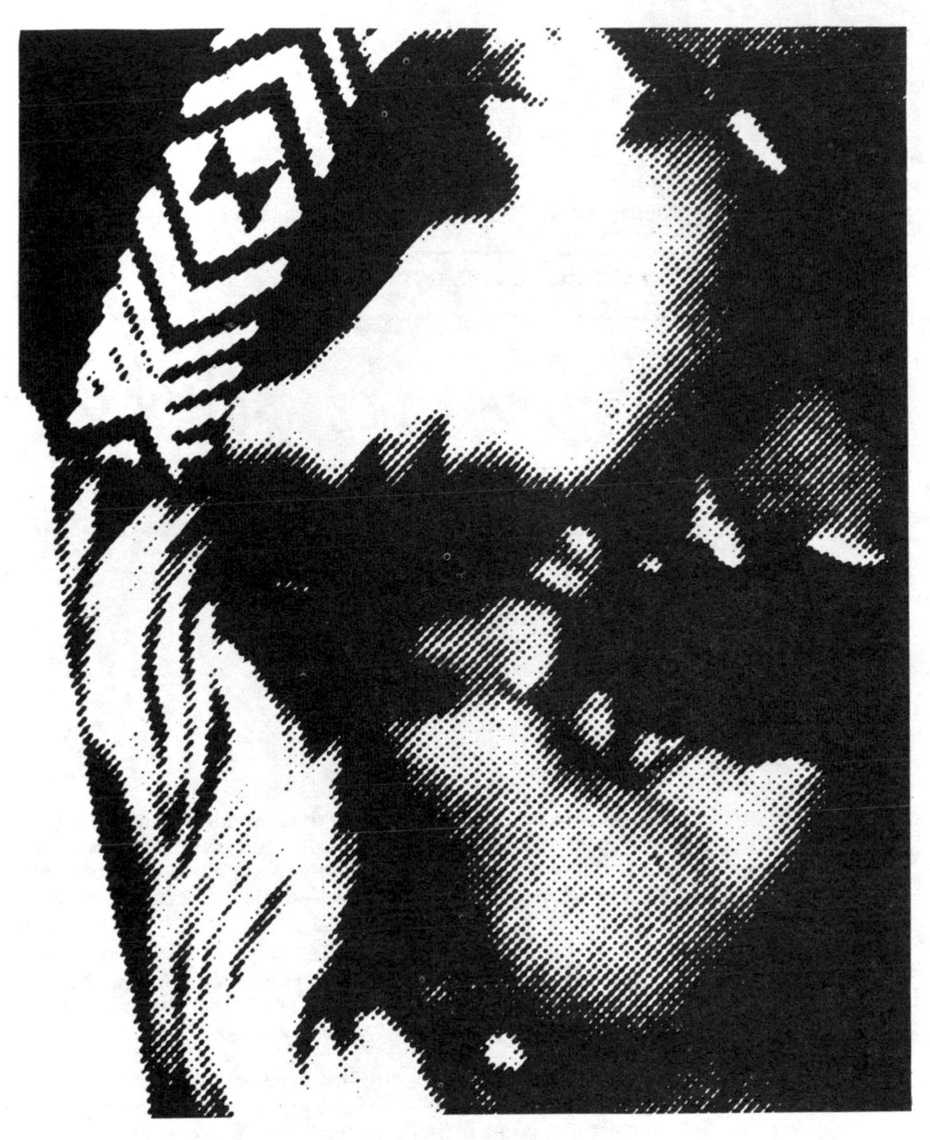

In video, a similar network of dots is graduated in terms of brightness, or 'luminance'. When the picture is first recorded, the video camera breaks it down into a fine mosaic of light and shade: white points of light where the picture is brightest, grey and black ones where it is dark.

Different video cameras work in different ways, and, with the introduction of new technology, the design of television sets and video monitors is changing. But, for the purposes of this broad description, let us look a typical use of conventional equipment.

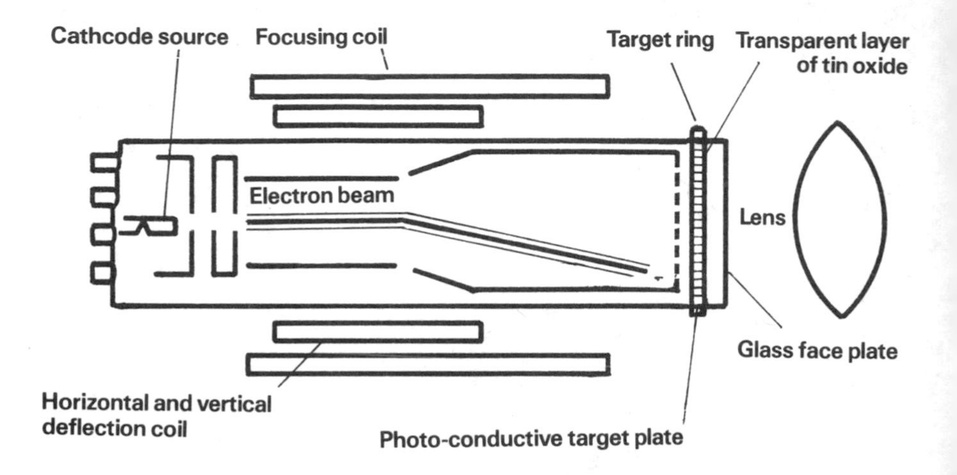

The vidicon tube in a video camera emits electrons from the cathode at the back of the tube. These are guided by magnetic deflection back and forth along the scan lines on the target plate. The resistance of the plate varies accordingly to the intensity of the light passing through any one point – and so does its complementary electrical output signal.

A 'target plate' near the front of the camera tube is coated with a photoconductive material (that is, one in which electrical conductivity varies according to exposure to light). This is the grid against which the mosaic pattern of light is laid. The brighter the point of light, the higher the positive electrical charge on any one spot.

This grid is 'scanned' by a narrow beam of electrons (negatively charged particles) emitted from a cathode at the back of the tube. (The cathode is the negatively charged conductor through which electric current enters the tube.) The negative charge of the electron beam 'discharges' the positive charge on each dot in the mosaic: the greater the positive charge on the dot (the brighter the point of light on that spot), the more electrons it absorbs.

Those electrons which survive this screen pass through to a 'signal plate', just behind the lens of the camera, which is coated with a transparent, conductive metallic film. Here, the electrons are used to generate a signal, the varying intensity of which reflects exactly the variations in the intensity of the light on the original mosaic grid.

A schematic diagram of a typical television receiver, showing the pattern of scan lines on the picture.

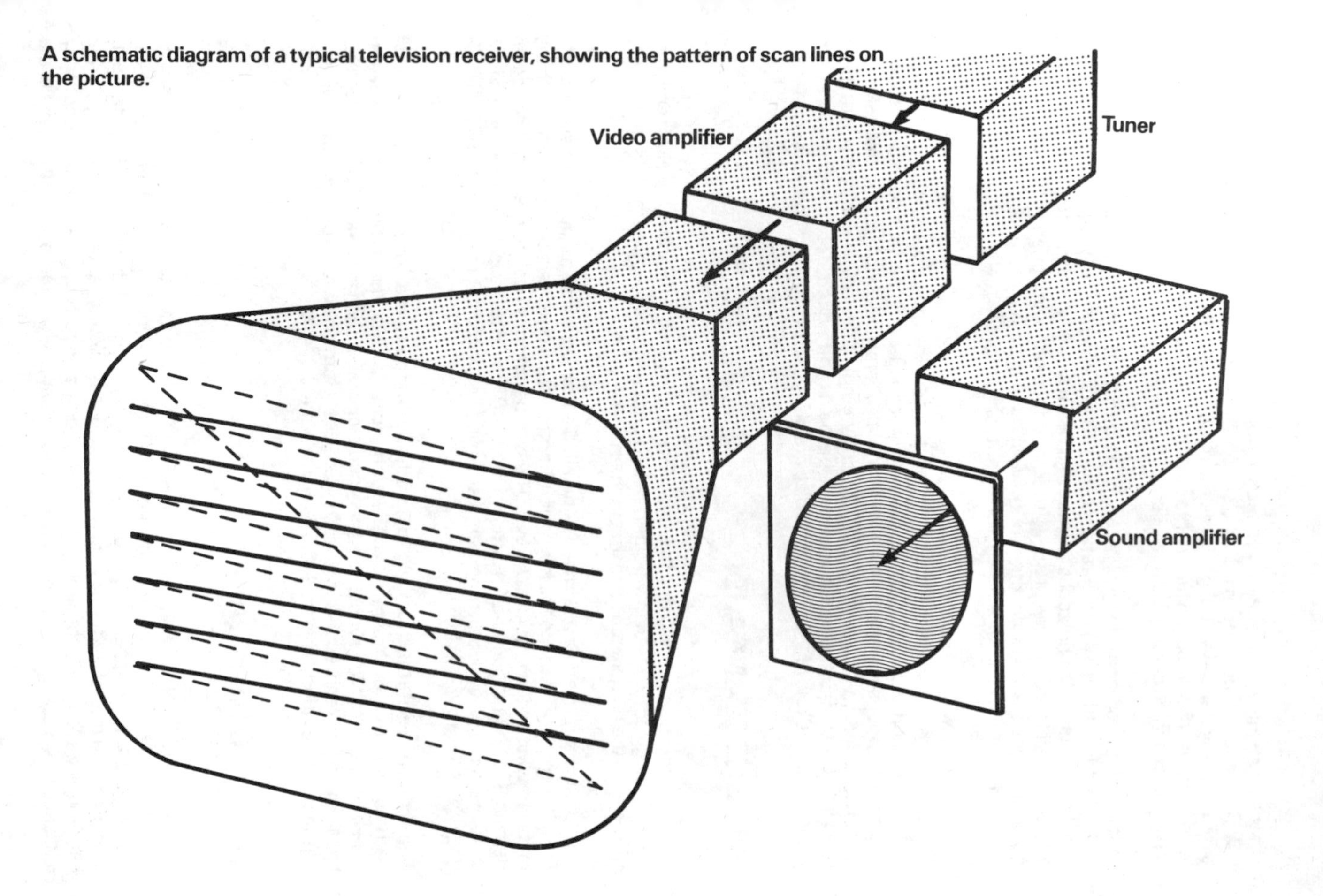

On the receiving end, the television receiver or monitor contains a cathode ray tube or CRT, a vacuum tube containing an electron-emitting cathode at its neck. (Properly, a CRT turns a signal into an image, while the camera tube that turns an image into a signal is an image orthicon tube.) The inner face of this picture tube is coated with a phosphor, a substance which can store signals sent to it as energy, and later release them as light.

In the CRT, the broadcast signal is converted back into a narrow beam of electrons (still of the varying intensity which reflects the pattern of light first recorded in the camera tube). Directed by the broadcast signal, this beam scans a grid on the phosphor-coated screen. Accelerated by a very high voltage, and focused by magnetic coils, the electrons build up kinetic energy on their journey. Where they hit the grid on the screen, it glows—the more electrons on any one spot, the brighter the point of light produced there.

This scan both translates the picture within the camera into an electrical signal suitable for transmission, and clears the target plate for the next image. Of course, this whole operation happens many times within a single second. (Most broadcast television and all non-broadcast video programmes currently employ a medium such as videotape to record this information, which can then be held, and edited, for transmission at any future date. The role of video tape and disc are discussed in the next chapters.)

THE LINE STANDARD

Naturally, for this scanning operation to work quickly and efficiently, the electron beam (or 'scanning spot') has to travel along a fixed path which is followed in the same way by the beam in the camera tube and that in the screen.

This path is arranged as a number of parallel 'scanning lines', running at a slight angle from the upper lefthand to the lower righthand corners of the screen. You can see these in the drawing of the television screen. The number of lines used to make up one screen describes the 'line standard'. Numbers from 405 to 819 have been used in various systems in the past, but the number of scan lines now commonly used are 525 (NTSC) and 625 (PAL and SECAM).

The scanning spot runs from left to right from the top to the bottom of the screen. Its path is directed by magnetic deflection coils within the tube, and its rhythm synchronised by a 'line sync pulse' recorded along with the video signal. The pattern of lines on the screen is called the 'raster'—the German word for 'screen', derived, descriptively enough, from the Latin word for 'rake'.

THE FIELD STANDARD

One more significant modification is needed to make this transmission work smoothly. Video, like film, is made up of many individual frames, a long string of snapshots taken uniformly at a fixed speed, each complete in itself, each slightly different from the last. (Film runs at a rate of 24 frames per second, video is slightly faster.) In slow motion, we can actually recognise these individual 'snapshots' but, run together at speed, they realistically simulate live action. As we have seen, when the eye and the brain are presented with a series of constantly changing images, moving too quickly to be identified individually, they interpret these as one coherently moving picture.

However, even if we could not distinguish single frames in these fleeting images, we would still be aware of light flickering on the screen as each one flashes on and off. But when the rate at which the frames are run is doubled, this flash is too quick to be seen, for 'flicker' is only detectable up to about 45 Hz. (An ordinary lightbulb, running on a mains power supply of 50 or 60 Hz, also flickers, but too subtly for us to perceive it consciously.)

In the cinema, this doubling is effected with the use of a multi-bladed shutter in the film projector, which shoots across in front of every frame, effectively splitting the time the image is seen on the screen into two equal parts. Therefore, each of the 24 frames is, in effect, seen twice separately within its allotted twenty-fourth of a second. The total effect is that of seeing 48 frames a second – and no conspicuous flicker.

In video, the picture is in fact scanned in two separate, interlaced passes. The first reads only the even-numbered scan lines, the second, only the odd-numbered ones. The lines in the first pass are 'double spaced', so the lines picked up in the second scan can be woven between them. As all this is happening within a fraction of a second, it's far too subtle for the eye to follow, and the effect is happily that of realistic, moving footage, without flicker.

FIELD SCANNING AND INTERLACING

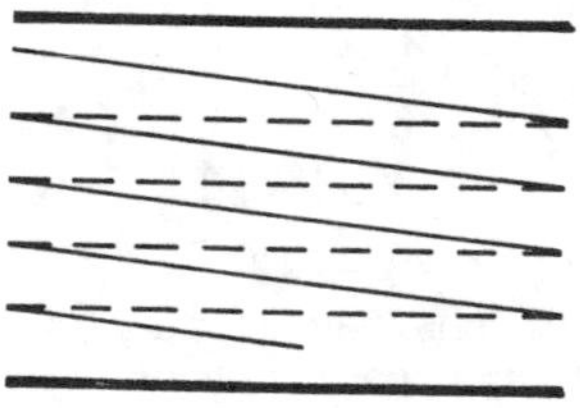

The first scan samples only half the lines on the screen: that is, one field.

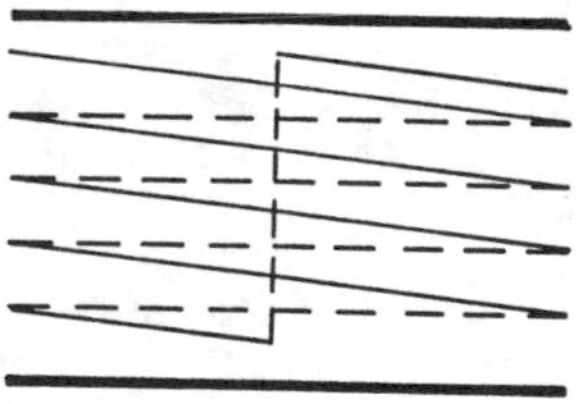

The second scan flies back upto the top of the screen to sample the remaining field.

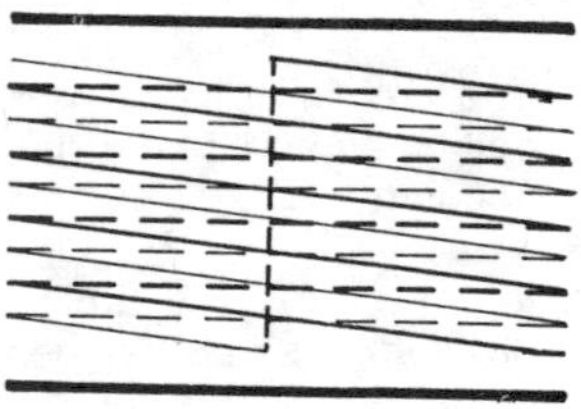

The complete frame comprises the two interlaced fields – and no discernible flicker.

These two, interlaced passes are called 'fields'. The rate at which they run (which is, quite simply, twice the rate at which the frames run) gives us the 'field standard'. This tends to be synchronised to the mains power supply frequency – 50 fields (or 25 frames) per second where the supply frequency is 50 Hz, but 60 fields (or 30 frames) per second where the mains power runs at 60Hz.

* NTSC runs at 60 fields (or 30 frames) a second.

* PAL and SECAM at 50 fields (or 25 frames) a second.

At the end of every field scan, the scanning spot shoots back from the lower righthand corner to the upper lefthand one, in a motion known as 'flyback'. This is synchronised by a 'field sync pulse', recorded on the videotape.

THE COLOUR STANDARD

So far, we have been discussing a black-and-white picture. As we mentioned, video, like the human eye, gathers information about light-and-shade (luminance) separately from that about colour ('chrominance').

The primary colours are red, green and blue (the beginning, the middle, and the end of the rainbow). These produce pure white when projected equally together, and combine in various degrees to form all the other colours we can see.

The luminance signal is always 30% red, 59% green and 11% blue: this is the colour combination that produces shades of black, grey and white. The chrominance signal on which the decoding of the whole colour picture is based does not send information about the colour picture as colour signals *per se,* but as colour difference signals: it explains how the true colour picture differs from that in the monochrome luminance signal.

The luminance signal modifies the chrominance one: it determines the 'hue', or tint, of a colour, and its depth, or 'saturation'. All this information is needed to present a colour accurately.

In all systems, the luminance signal is sent separately from the chrominance signal, so programmes can be seen in black- and-white on a system which cannot decode the colour signal. This is important not only in broadcast television, where some people have colour sets and some have black-and-white, but also, as we will see, in exchanging tapes between systems which have different colour standards.

NTSC is the standard introduced by the American National Television Standards Committee in the early 'fifties. It is also known as 'Never Twice the Same Colour': the slightest error in the phasing of the colour difference signals causes the receiver to decode the colours inaccurately, and apply too much of one or another to the final picture.

PAL (Phase Alternation, or Alternate, or Alternating, Line) is a refinement of NTSC. The signals are transmitted in the same way, but PAL uses more complex receivers to produce a more stable colour picture.

SECAM (Séquential couleur à mémoire–sequential colour with memory) is quite different. Whereas both NTSC and PAL transmit the separate components of the chrominance signal virtually simultaneously, in SECAM this information is sent sequentially, woven into the carrier signal. This makes the luminance signal more difficult to extract, so black-and-white reception of a colour-bearing signal is poorer than in NTSC or PAL. The pauses between the signals also make SECAM tape harder to edit than either NTSC or PAL.

BANDWIDTH AND FREQUENCIES

It is becoming clear that the video signal carries within it a great deal of information. There are, first of all, 25 or 30 frames a second, each containing 525 or 625 separate lines. In addition to the luminance and chrominance signals, there are, as we have seen, the regulating line sync and field sync pulses. There are also lines of pure black against which the luminance signal is graded. Then there are 'blanking lines' between each frame–this is where teletext information may be held. Altogether, some eleven million discrete elements are needed to construct a single frame of PAL standard video.

Video and audio signals are measured by the number of times the signal vibrates within one second (that is, the number of cycles per second). The standard unit for this is the Hertz (Hz), named after a 19th century German physicist; the range of frequencies that a video or audio signal uses is called the bandwidth. One Hz can hold two pieces of video information–so a signal containing eleven million pieces needs a bandwidth of five-and-a-half million Hz, or 5.5 MHz. (A single frame of NTSC standard video, having fewer lines, uses a narrower bandwidth–only 4.2 MHz.)

As we will see in the next chapters, finding ways of handling bandwidths of this size is a major problem in the development of any video system. Just consider that a high fidelity audio signal goes up to around 20 kHz, and that the human ear has a bandwidth of about 20,000 Hz, and you will begin to appreciate the nature of the problem. But more of that in a while.

COMPATIBILITY

Bandwidth is a problem for the manufacturers of video equipment. Video users face a more immediate obstacle: the incompatibility of the various standards. That is to say: an NTSC laser disc won't run on a PAL laser disc player, for the line, field and colour standards are all incompatible.

A SECAM tape will run on a PAL tape player–but only in black and white: the line and field standards are the same, but the colour standard is different. So far, only the VHD videodisc system handles NTSC, PAL and SECAM on one player.

The incompatibility of standards, once the exclusive concern of broadcast television, has become a headache for a lot of people who, a few years ago, wouldn't have known Betamax from Weetabix. Economically and socially, we depend on rapid transglobal telecommunications, and incompatible standards are barrier to the free exchange of information.

On a domestic level, you can't necessarily send a video tape to your relatives overseas as easily as you can an audio tape, even if you both have the same type of player. And the video equipment which looks like a bargain in a foreign market may well prove unusable back home.

And if incompatibility is a headache for home users, in corporate video applications, it can be a nightmare. A multinational corporation trying to implement its corporate video strategy across several countries faces a real dilemma: whether to choose locally-available equipment, and replicate the master tape accordingly, in a variety of formats, or to employ universally a system which may not be compatible with local standards (and which may, therefore, prove awkward and expensive to maintain and repair).

Arguably, the people who developed the television's first working prototypes could not have known what impact their work – and that of their contemporaries in quite different fields – was so soon to have on the fabric of modern life. The laying of the first transatlantic cable, and the debut of the first television receiver, hardly presaged in the popular mind the beginnings of a revolution which would sweep the world. Indeed, the people who developed those early models would have been considered visionary, to say the least, had they worked in certain anticipation of an international multi-media telecommunications network of the kind we take for granted today. The idea of the whole world being able to watch a broadcast live from the moon was the stuff of science fiction when John Logie Baird was unveiling his mechanical-scan television in the 1920s.

An article in a progressive home encyclopaedia of the mid-thirties describes a television broadcast over a distance of five miles in central London, and speculates that the introduction of cathode ray tubes may yet overcome the "obstacles to the realisation of television in a perfected, or even commercial, form". It goes on to describe Baird's "wireless receiving set" with its 7" × 3" screen – a machine which was displaced shortly thereafter, by EMI's system.[1]

We still have the problem of rival manufacturers backing incompatible systems. This opens a second, even larger, can of worms: the incompatiblity of formats - that is, the size, shape and operation of different systems. (So far, none of the popular videotape formats and players, and only some videodisc systems, are compatible with one another.) The incompatibility of standards affects all video production, whether for broadcast, corporate or domestic use, but the incompatibility of formats is an extra hurdle for non-broadcast users. This whole question is addressed more fully in the next chapters.

It is of course possible to reconcile material shot to different standards - using professional equipment in an editing suite. For instance, a 'telecine' converter (tele as in television, cine as in cinema) is used to reconcile film to video. Ironically, because of all these hurdles in video production, television programmes were once exchanged between incompatible broadcast systems on the universal medium of cine film.

There is a dreaded buzzphrase '3/2 pulldown' that describes the reconciliation of film running at 24 frames a second to NTSC video tape running at 30 frames (or 60 fields) a second. (This is where fields really come into play: every second frame of film is exposed on three rather than two fields, to produce sixty fields in every second's worth of tape. The end result is satisfactory, but not flawless: a still frame held on one of these extra fields will 'judder' badly.)

A simple solution in some situations is a dual- or tri- standard monitor or receiver, which can handle at least NTSC and PAL (and SECAM, if necessary). Some of these are quite compact, for corporate or social jetsetters.

Of course, various attempts have been made at electronic conversion systems, interpolating both lines and fields to reconcile material shot to different standards, and attempting to convert luminance and chrominance signals separately to translate colour pictures. The real breakthrough now seems to be, appropriately enough, computer-based digital technology, which breaks down video's analogue electrical signals into binary codes which can be fairly easily re-assembled into new configurations, without degradation of the picture quality. (Did you get all that? If not, refer to Chapter 18.)

Such solutions are costly and time-consuming—but some easy exchange of programming between different systems must eventually be found. With cable, pay-TV, round-the-clock broadcasting, and a proliferation of channels, broadcast television needs a vast number of programmes just to carry on from day to day, and the exchange of programmes from one network to another is one obvious solution to this problem. The exchange of material in non-broadcast applications is growing, too: good programmes are rarely cheap to produce, and big business as much as education appreciates the value of making one good programme and distributing it as widely as possible.

These differences could be swept away with the introduction of digital-based broadcast technology. It's difficult to imagine that anything other than a clean sweep would be possible to implement, if only for the cost involved. (It is sobering just to imagine how many private television sets there are in the world at this very moment.) But, until there is a change across the board, the incompatibility of standards is something with which a lot of video producers and users will simply have to be prepared to cope.

1 Anon. Home Management and Entertaining. Household Reference Library, London, undated. Page 589.

CHAPTER 11: HOW VIDEOTAPE WORKS

Videotape works along the same general lines as audio tape. Physically, it is usually composed of four bonded layers:

* The tape itself is made of polyester film, which is strong and flexible, but not elastic.

* A carbon backing reduces the build-up of static electricity when the tape is in use.

* A magnetically-sensitive emulsion, which commonly comprises a magnetic oxide powder (often iron oxide, although other metal compounds are also used), a binder and a lubricant, forms the recording surface.

* A neutral topcoat helps to protect the emulsion against dirt and damage.

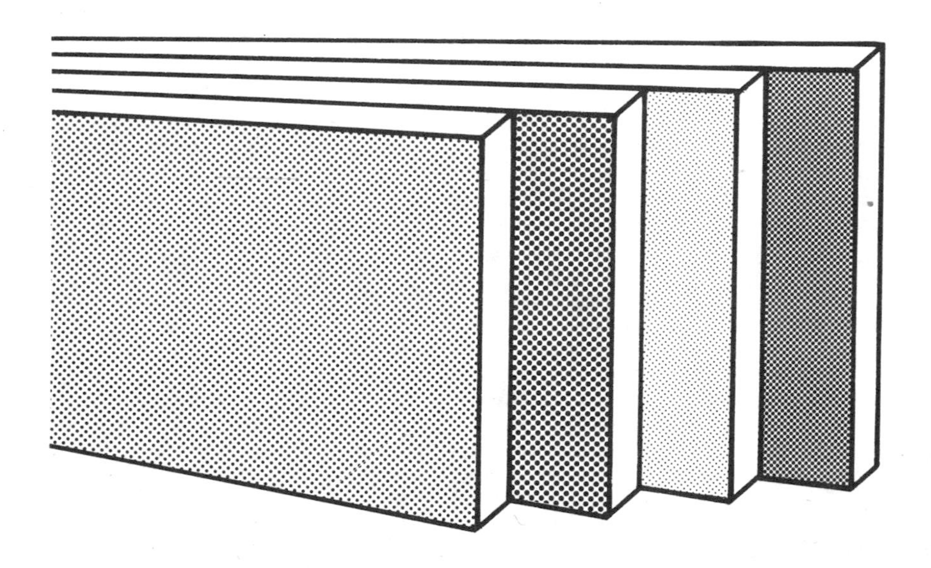

A smooth, clean oxide surface is essential; the pollution caused by dust and fingermarks impairs contact between the tape and the sensitive heads within the tape machine. This is why videotape is usually packed in cassettes which protect the tape from contact with anything but the very equipment that records and plays it.

Just as the videotape is usually made up of four layers, the information recorded on it is usually arranged in four parallel bands, or tracks, along its length.

* Audio signals are usually recorded in a narrow band along one edge of the tape, in closely-packed vertical tracks. (It is confusing, but the various signals within the four bands are also called tracks.) Audio signals are recorded and replayed by a separate audio head.

* A control track runs along a narrow band on the other edge. This is where the field sync pulse (described in the last chapter) is recorded; it regulates the running speed of the tape, just as sprocket holes control the running speed of cine film.

* A narrow cue track, which records signals, codes and verbal memoranda used in editing, often also appears along this edge.

* The video signal (and the line sync pulse) are recorded in shallow diagonal tracks on the wide band in the centre of the tape.

Narrow 'guard bands' separate the four tracks from one another.

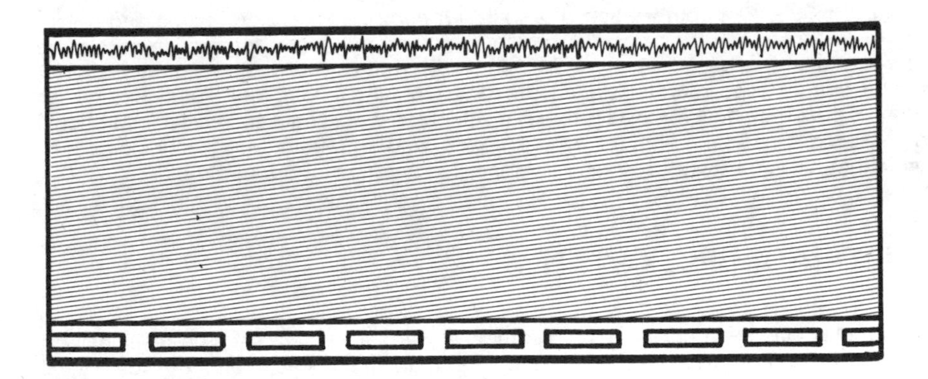

Typical arrangement of a helican scan videotape, with control and audio tracks on either side of the video track, and separated from it by guard bands.

MAGNETIC TAPE RECORDING

A similar principle is employed to record both audio and video tape. If you are, for example, an opera fan (and not too scrupulous about copyright infringement) you may once or twice have recorded a 'simul-cast'—that is, a broadcast transmitted simultaneously on television and on FM-radio—using both audio and video tape recorders. (This is rather cumbersome, but the best we have until the ultimate home entertainment centre efficiently combines all the possible audio and video media under one set of controls.) The two machines work in basically the same way.

The recording head on an audio tape machine is in effect an electromagnet with a slender gap between its poles. The magnetic flux across this gap is modulated by signals fed into the recording head. As the tape passes over it, the head rearranges the minute bar magnets in the metal oxide coating of the tape. At any given point on the tape, this configuration duplicates exactly the magnetic field around the head exactly as it was in the instant in which that particular track travelled across it. (Similarly, the erasing head simply shuffles all the particles in a random formation, ready for re-recording.)

When the tape is played back over the head (which in many machines is the same for both recording and playback), this variegated magnetic pattern induces fluctuations in the voltage, which are then amplified to reproduce the signals originally fed into the recording head.

This would give you a good stereo recording. However, just as there is more to opera than an audio recording alone can convey, so there is also more to video than to audio. As we saw in the last chapter, video signals need a much greater bandwidth—that is, a wider range of signal frequencies—than audio ones. While audio only needs to handle high fidelity signals up to about 20 thousand Hz, NTSC standard video systems need a bandwidth of 4.2 million Hz, and PAL standard systems, 5.5 MHz. Yet video and audio systems use tapes and equipment of much the same type and size.

The key to this is a combination of a high 'writing' (i.e. recording) speed and a very narrow gap between the poles of the recording head: the faster the writing speed, the narrower the head gap, the greater the bandwidth that can be recorded, and the better the quality of the recorded signal. Broadcast quality audio tape runs faster than domestic standard just for this reason. But videotape would have to move at improbably high speeds to work with the single, fixed head of audio tape technology. The breakthrough came in 1956, when the Ampex Corporation introduced Quad.

THE QUADRUPLEX SYSTEM

Quadruplex was the first videotape recording system, and is still used professionally. It takes its name from the fact that, instead of a single fixed head, it employs four heads mounted evenly around a rotating drum. Quad uses 2″ (50 mm) tape, which passes the drum at right angles to it. The width of the tape is drawn in a vacuum slightly more than a quarter of the way around the perimeter of the drum.

A schematic representation of a typical Quadruplex videotape, with audio, cue and control tracks, separated by guard bands, along the edges of the tape, and the wide video track in the centre.

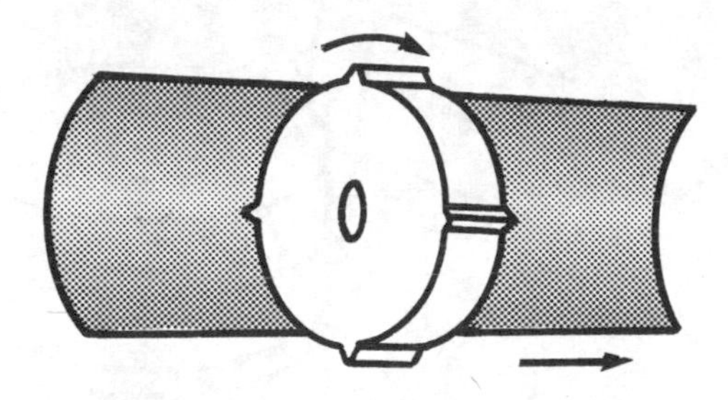

The rotating video head drum of a Quadruplex system, with the tape curving around the drum.

Each head writes one track from edge to edge across the tape, with a slight overlap so that each new track starts just as the one before it is finishing. Audio, cue and control tracks run along the length of the tape, with the video tracks, each holding about 16 scan lines, at a slight angle across the width (for this reason, Quad is also called 'transverse scanning'). The drum rotates at 240 rps in NTSC and 250 rps in PAL, and the tape moves at 15 inches per second (NTSC) or 15.625 ips (PAL) – about 35 cm/second.

The picture quality and editing facilities on Quad are excellent, but the system is too cumbersome and expensive for any but professional applications. So, Quad is being replaced, even in professional broadcast studios, by the tidier and cheaper 'helical scan' systems used in domestic standard videotape equipment.

HELICAL SCAN

Helical scan machines use two heads, mounted opposite one another on the video drum. When the player is loaded, the tape is drawn, lengthwise, most of the way around the drum, describing a helix: hence the name. This drum spins at 25 rps (PAL) or 30 rps (NTSC) – one revolution per frame – so in every rotation each head scans one field of a single video frame to produce one complete, interlaced picture at the end of every revolution.

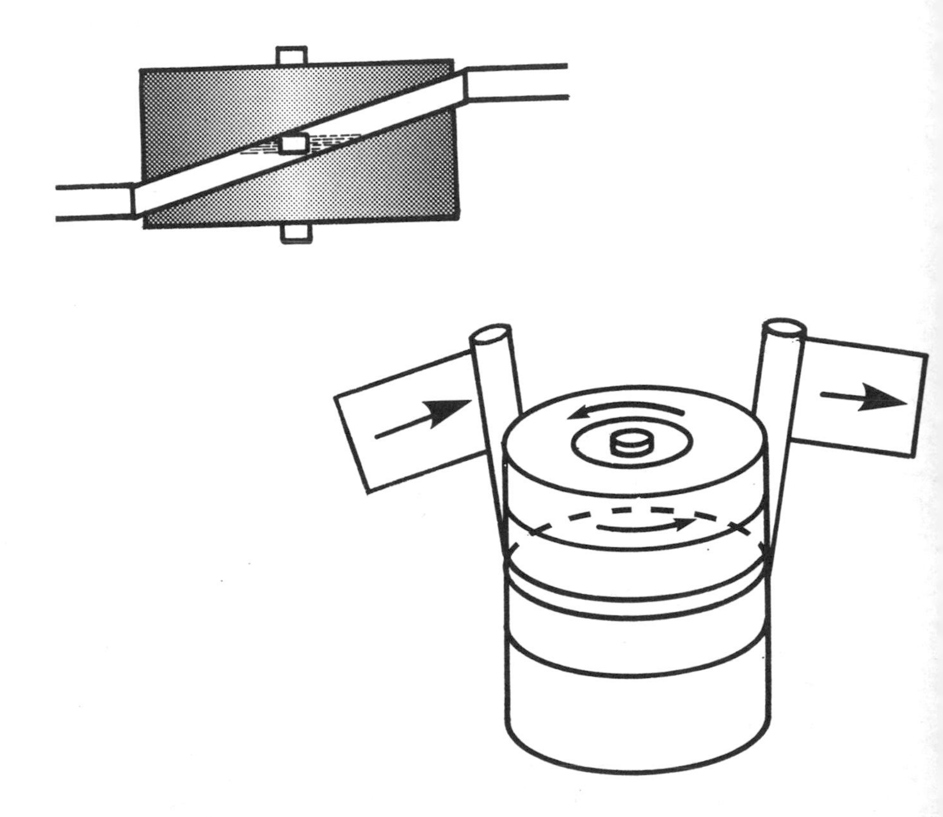

Helical scan systems take their name from the way the tape wraps around the video head drum, describing a helix. This schematic representation shows the path of the tape around the head drum, from the above and from the side.

As well as displacing Quad as a professional standard system, helical scan is used in both 'semi-professional' (or 'institutional') and 'domestic' (or 'consumer') formats, such as Sony's U-matic and Betamax, JVC's VHS and Philips' V2000 – all of which will be discussed in more detail in the next chapter.

WRITING SPEED

In both Quad and helical scan systems, the actual linear speed of the videotape passing through the machine is considerably less than its effective 'writing speed'–that is, the speed of the tape in relation to a head which is itself moving. For example:

* Betamax, Sony's consumer videotape format, has a linear speed of 1.873 centimetres per second, and a writing speed of 6.6 metres per second.

* JVC's VHS format has a linear speed of 2.339 cm/s, and a writing speed of 4.85 m/s.

* Quad, with a linear speed of about 35 cm/s (PAL runs slightly faster than NTSC), has a writing speed of about 40 m/s.

Thus, the use of multiple heads on a rotating drum effectively produces the high speeds needed to handle video's greater bandwidth without radically altering the technology refined and made familiar through audio tape systems.

TILTED AZIMUTH RECORDING

In helical scan systems, the drum is considerably wider than the tape itself, and is slightly tilted. Mounted in the machine, the tape drawn horizontally between parallel guide poles on either side of the video drum. But the drum itself is tilted, so the tape wraps along a sloping path around it. Thus, the video tracks which are arranged at an oblique angle to the edges of the tape are actually written and read (that is, recorded and played back) from start to finish horizontally when the tape is in its true position against the video drum.

The video tracks are densely packed (about a thousandth of an inch apart), with no room for guard tracks between them. Undesirable 'crosstalk' can occur when a head picks up signals from a track next to the one it is supposed to be reading. To minimise this, the two video heads on the drum are slightly tilted, and the information on adjacent video tracks recorded at complementary angles.

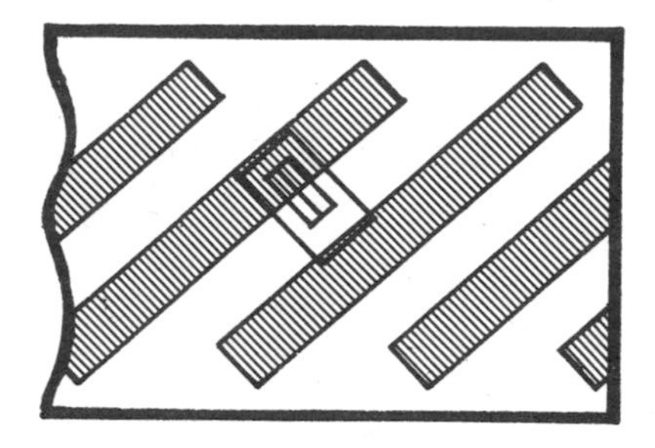
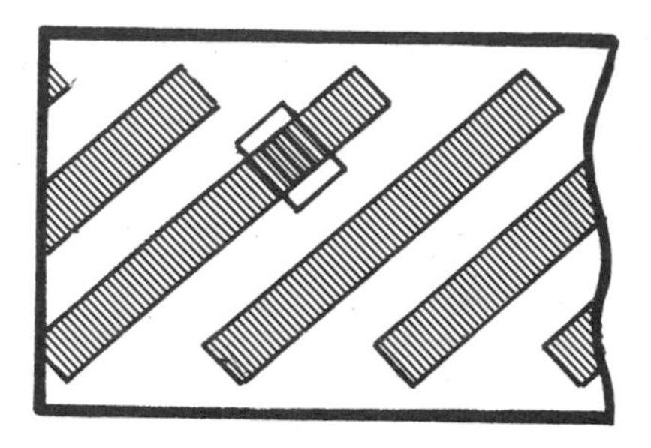

Early videotape systems needed 'guard bands' between each track of recorded material to prevent the reading heads from picking up signals from adjacent tracks—'crosstalk'.

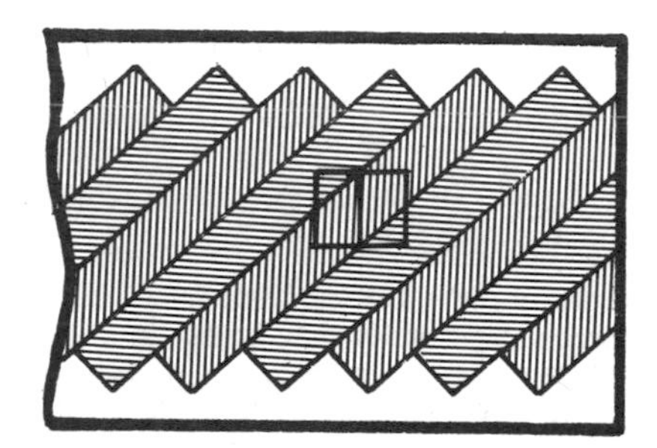
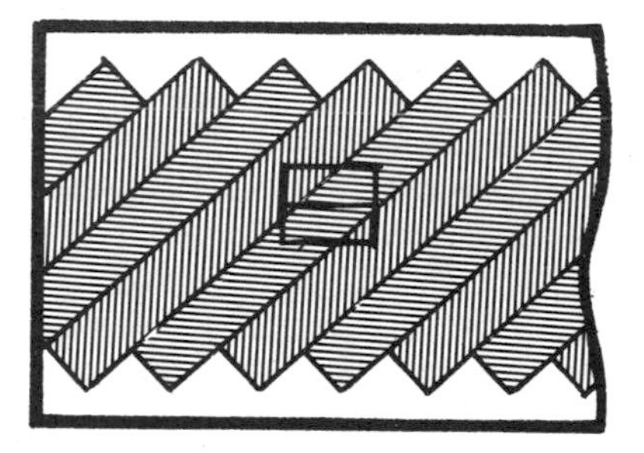

Slanted or tilted azimuth recording systems employ two video heads set at slightly varying angles, so that each reads only the track recorded at a complementary angle. This eliminates the need for guard bands, and increases the recording density.

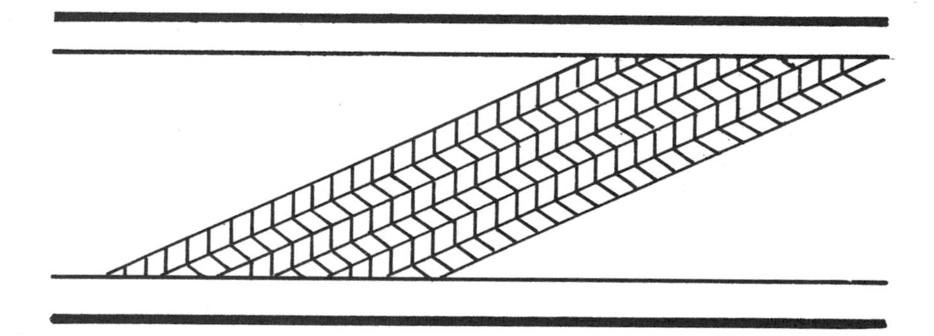

A schematic diagram of the 'herringbone' pattern of video tracks on a tilted azimuth recording.

The term 'azimuth' is borrowed from astronomy to describe the angular distance between the head on the circular drum and a theoretical north or south point on the rim of the drum: hence the terms 'tilted azimuth' or 'slant azimuth' recording. So, one video head is tilted 15° 'clockwise', the other, 15° 'counter-clockwise'. As you can see from the illustration, the pattern of the signals recorded on the adjacent tracks forms a sort of herringbone pattern.

Of course, the azimuth angle of the reading head must be exactly that of the recording head if the tape is to be replayed properly.

THE VIDEOTAPE CASSETTE

Open reels of cine film are a familiar image in popular art of all kinds. Professional quality one- and two-inch videotapes are also handled in single reels, but the narrower tape formats all come in cassettes which protect the delicate surface of the tape.

Sony's U-matic was the first cassette format. It was introduced in the USA in 1969, and in Europe in 1973, and is still in business, the only survivor of its generation. It was joined in the mid-seventies by the other formats which still hold the market – Sony's Betamax, JVC's VHS, and the several Philips' systems, for example. There are important differences between these, which we will explain in the next chapter. However, the adjacent diagram shows you what a typical cassette looks like within its sturdy shell.

THE VIDEOTAPE PLAYER

Videotape recorders and players also differ considerably from one another in their finer points. However, this diagram does illustrate a typical configuration of heads, rollers, capstans and guides described so briefly above.

The actual lacing of the videotape through this network varies with each system. This brings us on to the whole interesting subject of videotape formats, and to a fresh chapter.

The construction of a typical videotape cassette.

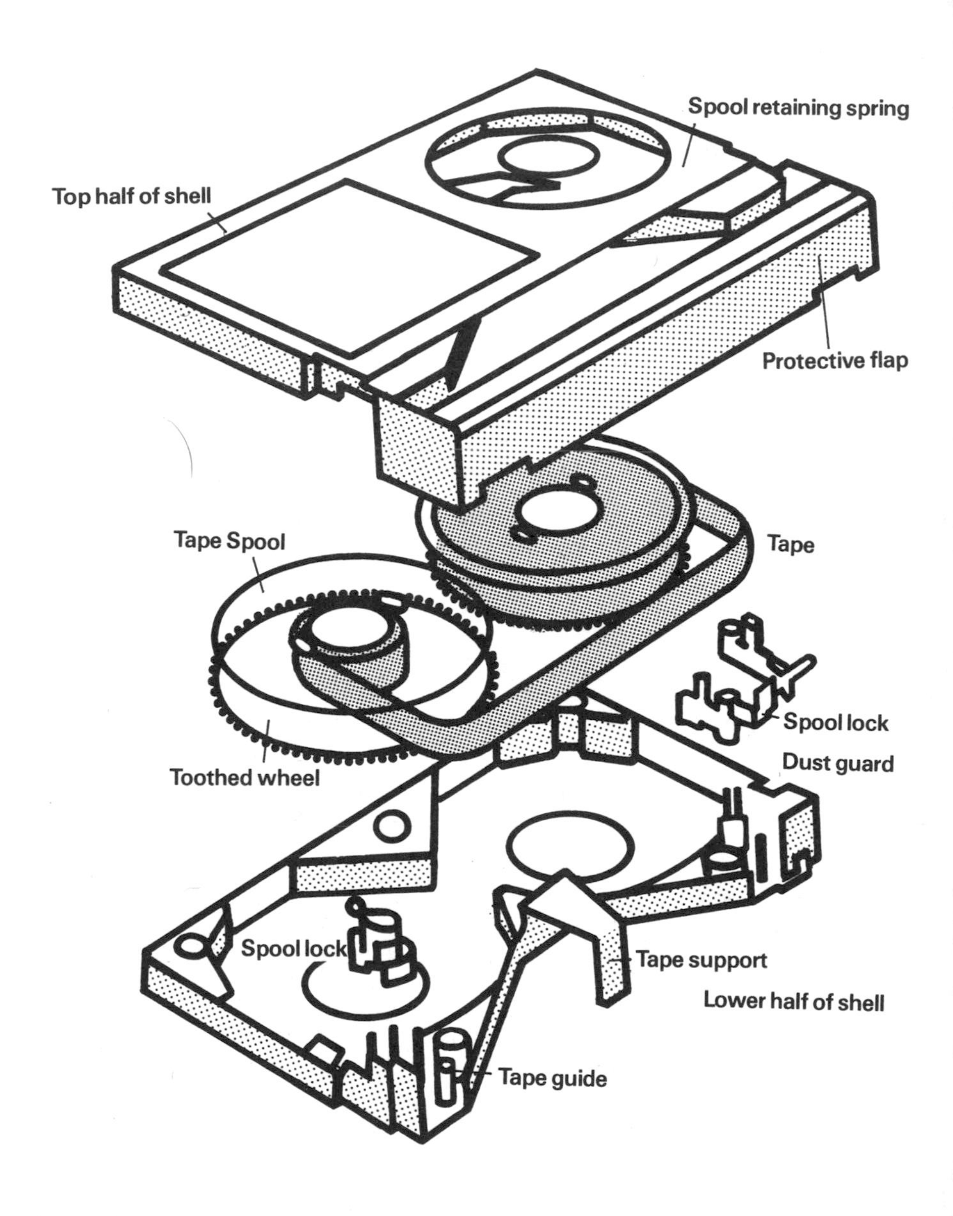

A video cassette

CHAPTER 12: VIDEOTAPE FORMATS

There are, first of all, five widths of videotape to consider. Some of these are offered in several incompatible formats, and together they span the current broadcast, professional, institutional and domestic markets.

TWO-INCH

Two-inch (50 mm) videotape is a high-quality professional format, used in broadcast television. It is handled in single reels rather than in the cassettes of the smaller formats. The Quadruplex system uses 2″ tape.

ONE-INCH

One inch (25 mm) tape is rapidly becoming the popular professional format, for it offers excellent quality, and is proportionally easier to handle than 2″ tape. It also comes in single reels.

Ampex, who developed Quadruplex, established what is known as 'A-format' in the 'sixties. A subsequent refinement of this, 'B-format', was backed by Philips, RCA, Bosch and IVC. The one most commonly used now is called, not surprisingly, 'C-format', and is produced by, for example, Ampex, Sony, RCA and Marconi.

U-MATIC (¾″): THE 'INSTITUTIONAL' STANDARD

Three-quarter inch (19 mm) tape represents the so-called semi-professional, or institutional, standard. There is only one format, Sony's U-matic (named for the U-shaped lacing pattern of the tape within the player.) The first videotape to adopt a cassette format, U-matic is still going strong, particularly in professional, non-broadcast applications.

In PAL standard systems, there is a choice between high-band and low-band U-matics. These are recorded using different signals, and require different players (a high-band colour tape will only play in black and white on a low-band machine). High-band tapes are broadcast quality, but low-band represents sufficient quality for non-broadcast users. This distinction does not exist in NTSC standard systems.

HALF-INCH: THE DOMESTIC STANDARD

Half-inch (12.65 mm) tape is used by two of the most popular domestic cassette formats, Sony's Betamax and JVC's VHS (Video Home System). Nevertheless, the two are incompatible: the Betamax cassette is appreciably smaller than the VHS cassette, and they lace in different ways.

Philips' V2000 (Video 2000), which replaces the earlier 1500 and 1700 systems, also uses ½" tape, but in a radically different way. The V2000 is the first system to record both sides of a videotape, by 'twin-tracking' so that the working surface of the tape is effectively ¼" (6 mm).

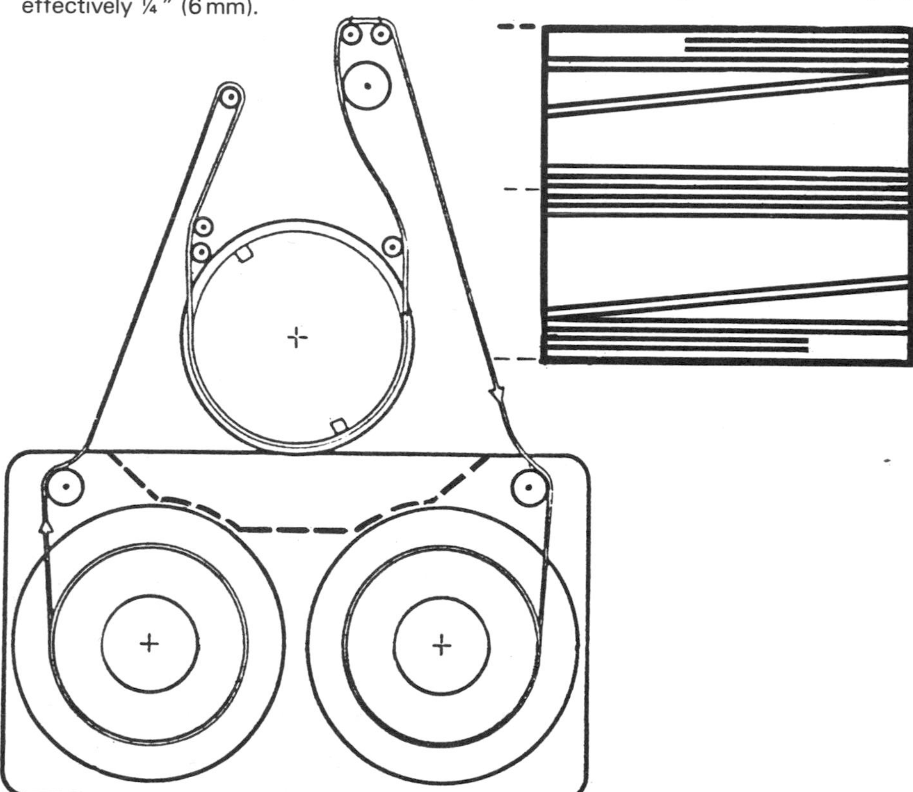

The lacing pattern and unique tape construction of Philips' V2000 videotape system.

This format is sometimes called VCC (Video Compact Cassette). Which is not to be confused with . . .

QUARTER-INCH: COMPACT FORMATS

Quarter-inch (6 mm) tape is the CVC (Compact Video Cassette) format. Also dubbed '8mm video', this compact format is a video cassette the size of a conventional audio cassette. One Japanese company, Funai, calls this format 'micro-video'.

Compact video is still very new, and it is interesting that over a hundred companies met in Tokyo to discuss the establishment of a common standard. Several independent projects were already well underway before any discussion of standardisation was attempted, so whether or when a common CVC format reaches the consumer market is another question.

BASF LVR AND TOSHIBA: LONGITUDINAL SCAN SYSTEMS

We have so far been discussing systems which employ the helical scan technology described in the last chapter. However, videotape systems have been developed which do not employ multiple heads on a rotating drum, but run along the same principles as audio tape recorders, using a single, fixed head. BASF even designed a system, LVR (Longitudinal Video Recorder), with the same running speed as audio tape, 4 metres per second; Toshiba developed a system running at 5.5 m/s.

The quad and helical scan systems were designed to overcome problems posed by the use of a fixed head. A videotape is stouter and shorter than the audiotape, and if the two are run at the same linear speed, the videotape plays through within minutes. The longitudinal scan systems answer this by recording both video and audio signals on a series of closely-packed parallel tracks running the length of the tape.

The Toshiba system uses 12.65 mm tape (the familiar ½" domestic size) on an 'endless loop'. With 300 parallel tracks recorded across 12.65 mm, a tape 135 m long provides two hours' uninterrupted playing time, with each tape loop lasting 24.5 seconds.

The BASF system works on the boustrophedon principle employed by Greek scriveners in ancient times and daisywheel printers in modern ones. The head automatically changes direction at the end of each track; 72 parallel tracks across an 8 mm wide tape, 600 m long, can provide three hours' uninterrupted playing time.

The longitudinal scan systems aren't nearly as popular as their rivals; however, their potential for miniaturisation is intriguing. The BASF cassette holds only one reel, which is threaded past a single video/audio head onto a spool on the opposite side of a central capstan. The system is already extremely compact, and could be made smaller still: perhaps small enough to fit straight into a hand-held video camera – a development which would considerably simplify location work. Interestingly, too, the large number of parallel tracks can all be dubbed – that is, re-recorded onto a duplicate tape – in a single pass along the length of the tape, another economical feature.

COMPATIBILITY

The salient point which emerges from all this is that no two videotape formats are compatible: even high- and low- band U-matics employ different colour systems, while Betamax and VHS are different sizes just for starters.

A brief comparison of the lacing patterns of, say, U-matic, Betamax and VHS within their respective players is illustration enough of the major structural differences which exist between different formats.

* Sony's two formats, the institutional U-matic and the domestic Betamax, follow a similar path, describing the eponymous 'U' shape around the video head drum. In both, the tape is laced when the cassette is inserted in the player, and stays in contact with the various heads, capstans and rollers until the cassette is removed from the machine.

* In JVC's domestic format, VHS, the tape is only laced when it is actually in use; the lacing describes an 'M' shape through the machine. To reduce strain and wear on the tape, functions such as fast forward are executed without engaging the video head.

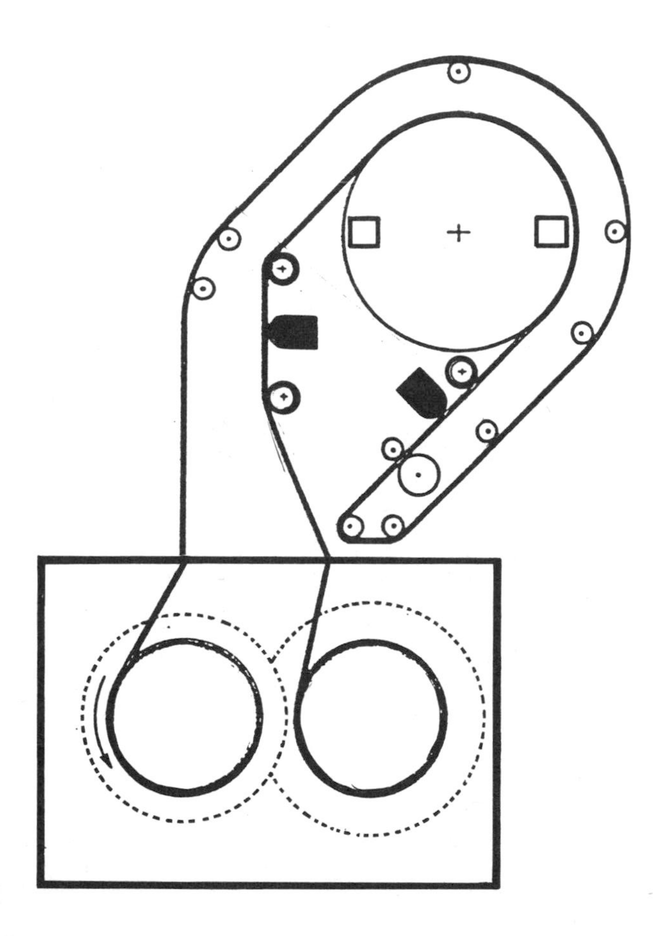

The lacing patterns of U-matic, Betamax and VHS videotape systems, respectively. These three all use tapes constructed along the lines of those illustrated in Chapter 11.

While the incompatibility of formats from rival manufacturers isn't exactly unprecedented, it is confusing and a little off-putting for many users, both domestic and professional.

The growth of the audio market, which has a much longer commercial history than video technology, is an interesting comparative study. Cylindrical tubes, and discs which ran from the centre to the outer edge, were tried and eclipsed in the earliest days of sound recording. Even after the ascendancy of the record as we know it, came the transition from 78 rpm to 45 and 33 1/3 in the 'fifties, the changeover from mono to stereo in the 'sixties, and the introduction of the compact, recordable cassette tape in the 'seventies.

In the 'eighties, the compact disc (CD) could mean the start of a new market, or a flash in the pan. Witness, in the collapse of the 8-track tape, one format which didn't have staying power, or, in the quadraphonic craze of the mid-seventies, an even more sobering lesson. (There was great rivalry to develop quad sound systems, and several were hurried into a market hyped but not totally sold on the idea. None was compatible with any other, and the whole thing promptly died a death. The second generation of these 'surround sound' systems is coming on better – and more slowly – but the tale is a cautionary one.)

However, the future of the video market as a whole does not concern us here, for not all videotape technology is suited to interactive video applications. That which is will be discussed in Section V. You could now go straight there, but we think that, whatever your ultimate plans, it would still be useful to read the intervening chapters, which discuss videodisc technology.

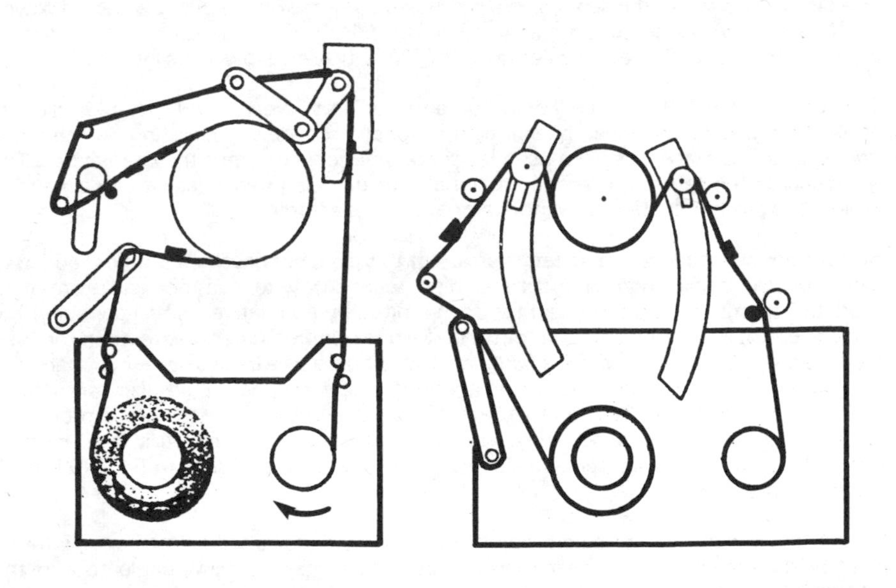

CHAPTER 13: AN INTRODUCTION TO VIDEODISC TECHNOLOGY

The very first videodisc was patented in Great Britain in 1927. It was developed by John Logie Baird, the man who, among many people working toward the same end, was the first to produce a working prototype of the television.

That set, the 'Televisor', was so dodgy that users could spend twenty minutes at a time fiddling about trying to intercept a signal. When the BBC broadcast for only an hour or two a day, that was no joke. So Baird developed a warm- up routine around a disc played on an ordinary gramophone linked to the television receiver. 'Phonovision', he called it. The disc was made of the same brittle plastic as contemporary audio discs, and played at 78 rpm to produce a dozen grainy still images. Both Phonovision and the Televisor became obsolete with the implementation of EMI's system a few years later.

The videodisc didn't surface again for another forty years, and when it did, its application was a specialised one. A rigid magnetic videodisc, operating on the same principle as videotape, has been used in sports broadcasting since the late 'sixties. The disc records the television broadcast signal, but can be interrupted at any time, and replayed to produce instant slow motion or even freeze frame.

Some work was done on a domestic version of this disc, but development has tended to focus on the application of magnetic disc technology to commercial information storage. The magnetic disc has so far been limited by its relatively small capacity: the storage density of a laser disc is something like one hundred times greater than that of a rigid magnetic disc, so to hold the same amount of data as a laser disc, a rigid magnetic disc would have to be about twelve feet across. 'Floppy' magnetic discs, similar to those used in computing, can be used to store still images. But while these are improving in quality and falling in price, they are still expensive and have limited storage capacity – 200 still frames per disc, compared to laser disc's 40,000 to 54,000 frames of moving and still footage.

In fact, the introduction of a commercial magnetic videodisc would further splinter an already divided market. The video disc has proven itself an invaluable tool in the industrial sphere, but the market is still unsettled, and there are several, incompatible systems competing for a share of it.

The first systems, Telefunken/Decca's Teledec (or TeD) and Matsushita's Visc, had a brief flutter in the mid-seventies and were heard no more. Then, in the late 'seventies, two commercially viable, incompatible systems appeared: the laser disc backed by Philips, Hitachi, Pioneer and Sony, and the CED (Capacitance Electronic Disc) system marketed by RCA as SelectaVision. In the early 'eighties, JVC (Japanese Victor Company) introduced a third system, VHD (variously decoded as Very High Density and Video Home Disc)–a capacitance system like CED, but not compatible with it. Laser disc, VHD and CED are discussed in the next chapters.

Among others, both the French telecommunications company Thomson-CSF and the American aerospace corporation McDonnell Douglas have developed systems of their own. (These are discussed in Chapter 16.) One commentator[1] estimated that in 1978, the year Philips launched LaserVision, there were forty different videodisc systems in various states of completion across the US.

It's interesting to see how many systems the various manufacturers have devised - even if, from a practical point of view, the laser disc and capacitance formats are the only two to make an impact on the commercial market so far.

TELEDEC/TeD AND VISC

The first videodisc on the consumer market was developed jointly by Britain's Decca and Germany's AEG-Telefunken. The Teledec, or TeD, system, with players and discs, was on the market in Europe in 1975–but not for long. Playing time was limited to ten minutes per disc and picture quality was poor. There were few discs to play once you had the machine. Later work improved the picture quality, but the system had by that time already died a death commercially.

In 1978, Matsushita (better known through its progeny, National Panasonic and Technics) launched a system called 'Visc'. It was similar to the TeD system, but offered up to two hours' colour video and stereo sound on a double-sided nine-inch disc.

Both systems employed the same principle as a conventional audio recording, with a stylus running along a spiral track. As we have seen in other chapters, the main obstacle to producing a video equivalent to the audio disc lay in the fact that a video disc would have to spin at a much greater speed than an audio disc in order to handle the wider bandwidth that video requires. (The laser disc, in fact, rotates around fifty times faster than an audio LP.)

A disc of conventional size, moving that quickly, would grind to pieces within minutes. The problem was to pack a much longer spiral track onto the disc. The big breakthrough was that of changing the plane of the undulations in the track from lateral to vertical.

The track in an audio disc is like a shallow river, flat- bottomed, and meandering from side to side. That in a video disc is more like a canyon, with sheer sides and a rocky floor. To get an hour's playing time out of one side of a disc that spins at 1500 or 1800 rpm, the spiral track must be extremely long and closely packed.

The TeD and Visc systems both employed a prow-shaped stylus equipped with a piezoelectric transducer (similar to, but smaller than, the familiar crystal pickup of an inexpensive record player). The TeD disc was flexible, made of flimsy plastic foil; the Visc was made of the same rigid vinyl as an ordinary LP. The TeD system rotated at 1500 rpm (PAL) and 1800 rpm (NTSC), the Visc at 375 rpm (PAL) and 450 rpm (NTSC). However, neither was the right product at the right price for the market of the time. The lack of programming – that is, discs to play on the machines – undermined the success of both. Eventually, Matsushita deferred to its sister company, JVC, who were developing quite another system, based on the capacitance principle; the fruit of their work is discussed in Chapter 15.

Meanwhile, it was another system altogether which really seized the popular imagination: the laser disc. This is discussed in the next chapter.

1 Hope, Adrian. 'Disc Players'. Newnes Book of Video, K G Jackson, ed. Newnes Technical Press, London. Page 42.

CHAPTER 14: VIDEODISC FORMATS: LASERDISCS

The reflective optical disc, familiarly known as the 'laser' disc, looks like a fancy dress version of the conventional audio LP—slightly thicker and heavier, and with the silvery prismatic reflective coating so effective in colour advertisements. Unlike an audio disc, its surface is smooth, for the essential information is encoded well beneath the tough protective coating.

Laser disc is a sub-set of the optical disc technology discussed in Chapter 9 and again in Chapter 17. For our purposes, the essential difference is that optical discs can record data of all kinds, from video signals to oil well production records, whereas videodiscs are so far dedicated to sound, pictures and the occasional computer program.

MANUFACTURING

This description is based on Philips' manufacturing process, which represents laser disc technology in the hands of the company identified with its commercial inception. The laser disc as we now know it was developed by Philips and MCA. However, to avoid rivalry of the kind which, for example, brought the first quad sound systems to ruin, other manufacturers (notably Hitachi, Pioneer and Sony) have been encouraged to adopt the format. Their systems differ from this in some details; the compatibility of laser discs as a whole is discussed later.

The 'master edit'—the final assembly of video and audio material—is made on videotape. It first goes through a stage called 'pre-mastering', to ensure that it is properly prepared for the 'mastering' stage to follow.

The master disc is made of glass, optically ground and polished. A 0.1 micron membrane of light sensitive 'photoresist' is laid over this to form the 'substrate', or foundation, for the recording stage, which is essentially photographic. The master disc is 'cut' by a laser beam which incises a pattern in the photoresist membrane, modulated by the signals recorded in the master tape. This is developed photographically to produce the pattern of shallow pits along a spiral track which is the heart of the reflective optical disc system.

The disc that actually reaches the consumer market is several generations removed from this fragile glass master. The generations which follow are in fact called Father, Mother and Son. Following this genealogical metaphor, the discs which are eventually turned out in quantity are great-grandchildren of the glass master, and great-great-grandchildren of the master edit.

The freshly-cut master disc is first silvered (both for conductivity, and also to facilitate inspection) and then electroplated with nickel. This metal plate is then separated from the glass master, carrying away a negative impression of it. This is the 'Father'. The negative profile is required for the manufacturing stage, but this first impression is still too fragile to be used to stamp out any quantity of replicas. So more discs are produced, in patriarchal order, each member of the family plated off the one before: the positive profile of the Mothers off the negative profile of the Father, and the negative profile of the Sons off the positive profile of the Mothers. The single Father can produce many identical Mothers and Sons. The Sons of this curiously incestuous family are used to stamp out the final, marketable discs.

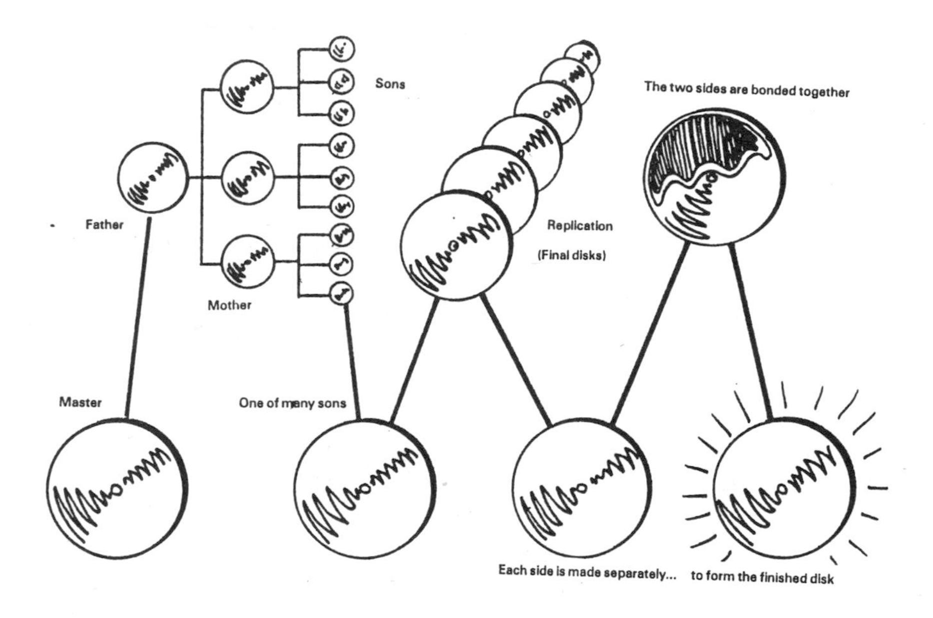

Stages in the manufacture of a laser disc.

The basis of Philips' manufacturing process is a transparent prefabricated disc, made of polymethyl methacrylate (PMMA), a plastic more familiarly known as Perspex or Plexiglass. The final disc is essentially built from the top down. One side of the PMMA layer remains smooth, and forms the outer surface of the finished disc. The other side is stamped from the negative profile of the Son, to produce a positive profile descended from that of the glass master—much in the way that an audio disc (or indeed a coin or a medallion) is stamped from a mould. This replica is immediately coated with a thin layer of special lacquer. An extremely fine aluminium membrane is laid over this, to produce the essential reflective layer. A protective plastic coating completes one single-sided disc.

All discs are double-sided, whether or not there is anything recorded on the second side. So two single surfaces are bonded together with the smooth face of the PMMA layer forming the outer shell.

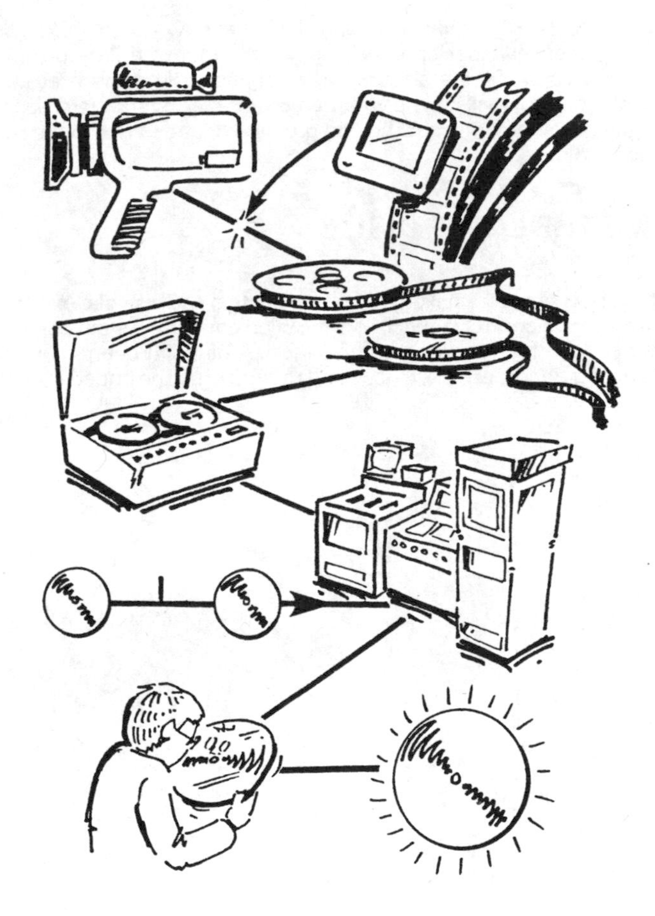

An optimist's impression of the work involved in producing an interactive videodisc.

There are different ways of pressing the ultimate plastic disc, but the basics of the recording and manufacturing process are pretty much the same for all the videodisc formats currently on the market. Not, of course, that this is the last word in videodisc manufacture.

In America, Quixote Corporation have devised a manufacturing process which is apparently more economical of money, time and materials than established methods. The master disc is glass coated with metal, and records signals as holes rather than bumps or pits and grooves. The body of the final disc is PMMA coated with reflective metal. The master is laid against it, and the two are exposed to light which passes through the holes in the master onto the face of the replicate. The replicate is then processed to develop the pattern of holes in its reflective surface – a process not unlike that by which semiconductors are made. The system can also be used to manufacture digital audio discs.

In Japan, Matsushita are moving toward high-density storage. Established laser disc technology, figuratively speaking, records on the floor of a flat-bottomed canyon; this new process records on the sides of a V-shaped valley, so an even narrower (and, consequently, longer) track can be wrapped onto a disc of standard size. Both processes are still rogue developments, but they do presage changes which are bound to come in so new a technology.

PLAYING THE LASER DISC

In play, the disc is read by a helium neon laser within the disc player. The beam is locked on track by a servo- control (a mechanism which converts a small force into a large one) to maintain precise focus and synchronisation. A pinpoint beam of intense light is sent along a path of gratings, prisms, lenses and mirrors onto the underside of the disc.

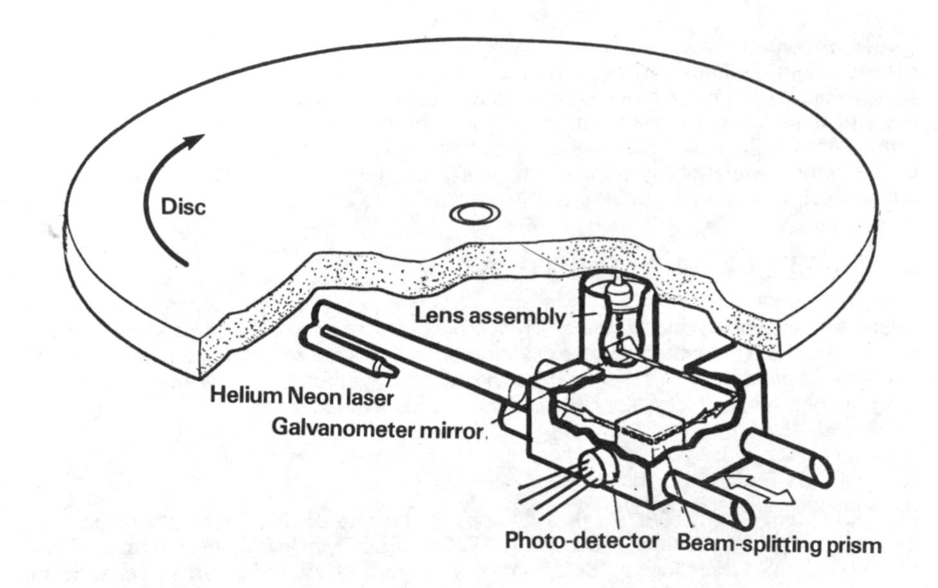

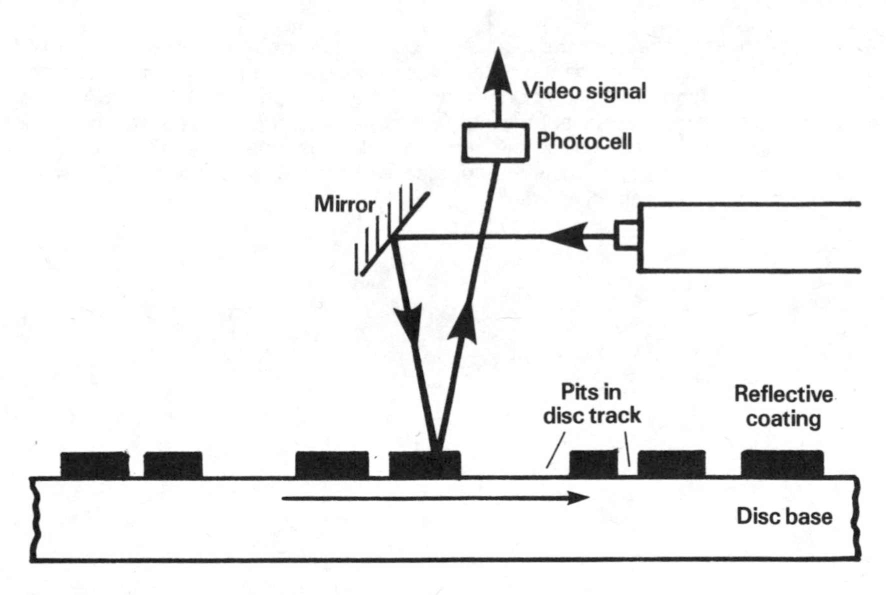

The inner workings of a Philips LaserVision videodisc player.

Under the control of a scanning lens, this light hits the surface of the disc and is bounced off the reflective aluminium layer. This is of course a precise imprint of that pattern of shallow pits and grooves first recorded on the glass master disc. The uneven pattern of the pits and grooves causes variations in the light reflected off the surface of the disc. The reflected light is sent back along the path and through a photosensitive diode which converts these variations into the electrical signals from which video and audio signals are derived.

ACTIVE PLAY AND LONG PLAY

There are two types of disc within the reflective optical system: CAV (Constant Angular Velocity) and CLV (Constant Linear Velocity). In a videodisc catalogue, you will see that CAV (or 'Active Play') discs feature prominently among, for example, exercise programmes, while CLV (or 'Long Play') discs are reserved for movies.

CAV ('Active Play') Discs

CAV (Constant Angular Velocity) discs spin at the rate of one frame per revolution, which is a constant speed of 1800 rpm in NTSC, or 1500 rpm in PAL. (Remember, NTSC runs at 30 frames a second, or 1800 frames a minute, and PAL at 25 frames a second, or 1500 frames a minute.) CAV discs now offer about half-an-hour's straight playing time per side, depending on who made the disc. (Philips' disc holds 54,000 frames per side, which is 30 minutes' linear playing time per side in NTSC, 36 minutes' in PAL.) Of course, in an interactive programme, actual 'use time' could be more or less than this, depending on the application and the user.

The track length – the space on the disc assigned to each separate frame – is not fixed, but is, obviously, much longer on the outer edge of the disc than on the inner. The neat ratio of one revolution of the disc to one frame of video material means that CAV discs are what is known as 'frame addressable' – that is, any single frame can be easily identified and quickly and accurately retrieved. This is one component of disc's impressive 'random access'. Precision, coupled with speed, puts the CAV laser disc ahead as a medium for interactive video.

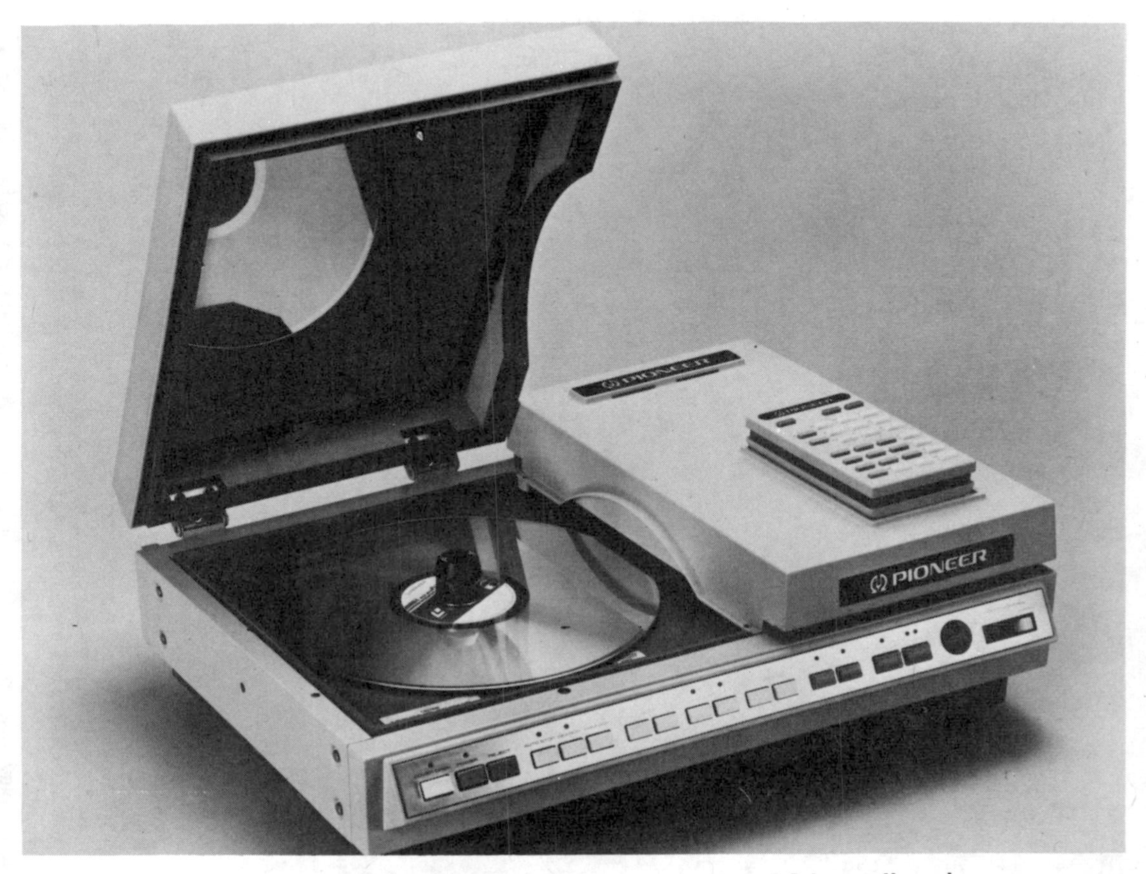

Pioneer's early NTSC-standard industrial quality Level 2 laser disc player, the PR 7820.

The VP835 PAL-standard 'Professional LaserVision' videodisc player from Philips. It carries an RS-232 computer port, and among its optional features includes a 48K onboard microprocessor that takes plug-in EPROM cartridges.

CLV ('Long Play') Discs

CLV (Constant Linear Velocity) discs, on the other hand, assign a fixed track length to every frame in the programme, so there are more frames read in one revolution around the outer edge of the disc than the inner. The disc does not play at a constant speed, as the CAV disc does, but runs more slowly on the rim than near the centre. This greatly increases the playing time: Philips' CLV disc, for instance, can hold 87,000 frames a side, which gives nearly two hours' playing time on a double-sided disc. However, this also reduces access from the frame accurate address of CAV discs to terms of chronological time, minutes and seconds. Random access is, therefore, much less precise on these 'extended play' discs than on 'active play' ones. The short answer is that CLV discs are an excellent medium for entertainment, but of limited use interactively.

COMPATIBILITY

Theoretically, laser discs made by rival manufacturers to the same international standard (NTSC or PAL, described in Chapter 10) can be used interchangeably. This is certainly true at the simplest, linear level – that is, when the disc is playing from start to finish without interruption (Level 0 on the scale outlined in Chapter 5). It is also the case at the top of the line, at Level 3, when the disc player is working under the control of an external computer.

This means movies and the like can be played interchangeably on different laser disc players, as can interactive video programmes employing the same outboard computer. However, the onboard microprocessors of the rivals' various players do differ, so a disc functioning at Level 2 cannot always be used on a player other than the one for which it was made. A laser disc with all its programme instructions actually encoded on the disc for use with one particular disc player would not work interactively on a completely different player. (It would, of course, still play linearly or under external computer-control at Levels 0 and 3.)

However, a disc made for either of the capacitance systems described in the next chapter won't play on a laser disc system. (Nor even on a different capacitance system: there are two, and they are incompatible). But this is, aside from a cautionary note, the stuff of the next chapter.

CARING FOR YOUR VIDEODISCS

Dust and grime on the surface of the disc won't harm it, but the laser disc should still be handled carefully.

Grubby fingermarks generally do not affect the disc. It is relatively indifferent to scratches and other such surface damage—but it is not quite so rugged as early enthusiasts once suggested. The outer face of the disc does protect the material beneath, but deep scratches (especially into the control tracks) can damage the disc irreparably.

Altogether, it is a good idea to handle a videodisc as carefully as an audio LP, for it is not indestructible. In fact, if dropped at a certain critical angle, it will shatter as finely as the Snow Queen's mirror. (We only tell you this: we are not saying at which angle, nor do we recommend that you experiment to discover it for yourself.)

THE VIDEODISC MARKET

From a commercial prospective, it seems the first years of all the videodisc systems have been fraught with setbacks and delays. RCA were some time getting their capacitance system, SelectaVision, onto the American market, and even then sales were slow. Yet one commentator noted that compared to sales of colour television receivers in 1954, and video cassette recorders in 1976, SelectaVision's sales in its first year were in fact encouraging.[1]

By 1972, MCA and Philips were both formally committed to the development of the laser disc, and agreed common standards. MCA formed DiscoVision Associates (DVA) with the ubiquitous IBM. DVA had a high profile in disc's early days, but was wound down in 1982 and its effects—including a pressing plant in Costa Mesa, California—sold to Pioneer, who immediately revamped the whole organisation, installing top-class Japanese equipment, cutting down the work force and improving production dramatically. In the meantime, Philips changed their disc's name from VLP (Video Long Player) to LaserVision. By the late 'seventies, laser discs were firmly established on the American and Japanese markets.

The move into the European, PAL standard, market came next. Philips were the first in with consumer players. Pioneer followed, with a more versatile consumer player. Philips then began to introduce their PAL standard 'professional' players. Before there had been PAL standard equipment, some projects used imported Pioneer NTSC industrial players.

Meanwhile, various other systems were progressing fitfully. The debut of JVC's capacitance system, VHD, was on-again, off-again throughout 1982, while Thomson-CSF and McDonnell Douglas, perhaps temporarily, withdrew from the race. VHD's debut finally took place in Japan in the spring of 1983, and in the UK that autumn, with programming from JVC's European software partners, Thorn-EMI. That same autumn, RCA unveiled the first fully interactive CED player in the USA, and brought the CED system in its simpler form to the UK.

Much of the commentary on videodisc's commercial potential is based on the reactions of the consumer market. This is misleading, for the disc, as a medium for interactive video, has established a place for itself in professional applications which no other format has yet touched.

Many domestic users feel disc won't rival tape until it can record easily and cheaply at home. That time isn't far off. But, even now, many professional users feel that, for interactive video especially, the disc has left tape standing. In truth, there is room for tape and disc – and for interactive tape/slide, too. Whatever its ultimate position in the jockeying for the consumer market, the videodisc is certainly an established and flourishing part of the world of professional video.

1 Screen Digest, June, 1982, p 112:

'The actual sales (a little over 100,000 players in the first year) are four times the 25,000 colour television receivers sold in 1954 . . . and more than three times the 30,000 video-cassette recorders . . . sold in 1976.'

CHAPTER 15: VIDEODISC FORMATS: CAPACITANCE SYSTEMS

There are currently three, incompatible systems competing for the videodisc market. We looked at reflective optical discs (laser discs) in the last chapter. Now we turn to capacitance (or 'contact') discs.

There are two, incompatible capacitance systems – that backed by RCA, and that of the Japanese Victor Company (JVC) and its European software partner, Thorn EMI.

* JVC's system is called VHD (Video High Density).

* RCA's system is called CED (Capacitance Electronic Disc), and is marketed under the name SelectaVision.

In a nutshell, 'capacitance' means the capacity to store an electric charge. We won't go into the details of volts and farads and electrical potential in these pages: you can read a book on electronics if you're that way inclined. The salient point is that capacitance discs handle audio and video signals entirely by electrical means, through the agency of an electrode-bearing stylus or sensor actually in contact with the surface of the disc. (Laser disc systems use variations in reflected light to decode recorded information: there is no physical contact between the laser disc and the device that reads it.)

Capacitance discs look like a cross between conventional audio LPs and computing's floppy diskettes. The tracks on a videodisc are packed at least forty times more densely than those of an audio LP, and the naked disc's tolerance to dirt and wear is low. This is why laser discs are recorded beneath a tough, translucent plastic coat. Because there must be direct contact between the disc and its reading head, capacitance discs forfeit this coat and are instead housed in rigid plastic jackets ('cases' or 'caddies') which they leave only within the shelter of the player. The whole case is loaded into the player, and the disc extracted only there, out of harm's way.

The discs are either made from, or coated with, an electrically conductive plastic. Like laser discs, they store information about video and audio signals as a pattern of shallow pits recorded along a spiral track on the surface of the disc. Like audio discs, they employ a diamond or sapphire stylus or sensor which carries a conductive metal electrode. Differences in capacitance between the surface of the disc and the electrode in the stylus are taken up by a tuned circuit to produce video and audio signals.

CED – THE GROOVED CAPACITANCE DISC

The RCA system – CED, or SelectaVision – is called the 'grooved' capacitance disc because actually has a shallow groove to direct the stylus in its journey. However, it is not mechanical, as the stylus of a stereo cartridge is. Rather, the electrode-bearing sapphire stylus senses changes in the electrical capacitance of the metal elements in the disc. The disc rotates at 450 rpm for NTSC and 375 rpm for PAL; it currently offers about two hours' playing time on a double-sided disc. Each revolution holds four frames.

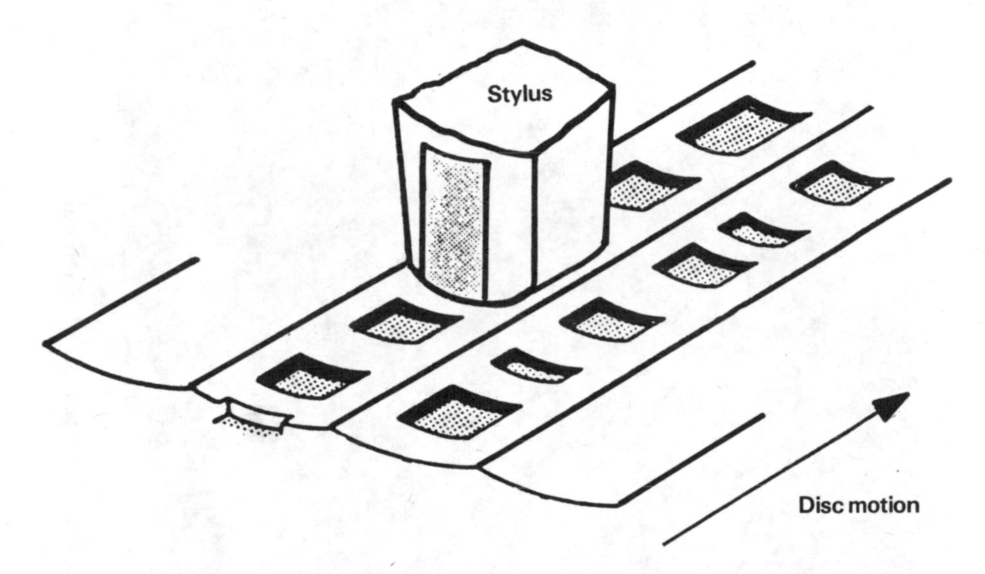

The CED, or 'grooved' capacitance videodisc developed by RCA.

We explained in the last chapter that CAV laser discs' ratio of one frame per revolution facilitates accurate random access and freeze frame. With four frames per revolution, CED systems cannot yet freeze on any frame at random, as laser discs can, but have to record still images on four (or, preferably, twelve) separate, identical frames, to ensure stable and accurate access. Laser discs did this in their early days, too, so the problem may yet be overcome.

Initially, CED appealed chiefly to the home entertainment market – it is low-priced, attractive and easy to operate, with a good choice of programming. (For an interesting perspective on CED's early sales performance, see the footnote to Chapter 14.) However, in the autumn of 1983, RCA introduced a Level 2 interactive player in the US which, while less versatile than contemporary laser disc systems, is certainly equal to the demands of many applications – and undercut the prices of laser disc players by up to 60%.

The CED capacitance videodisc player from Hitachi.

VHD – THE GROOVELESS CAPACITANCE DISC

The JVC system, VHD, is grooveless: the spiral track of 'programme pits' containing the video and audio signals is interlaced with one of 'tracking pits' which guide a sensor over the smooth surface of the disc. The disc holds 45,000 still frames and spins at 750 rpm (NTSC) or 900 rpm (PAL), a rate of two frames per revolution. VHD has a linear playing time of two hours per double-sided disc, with the basic features of interactivity – the best of CLV and CAV laser discs in one format.

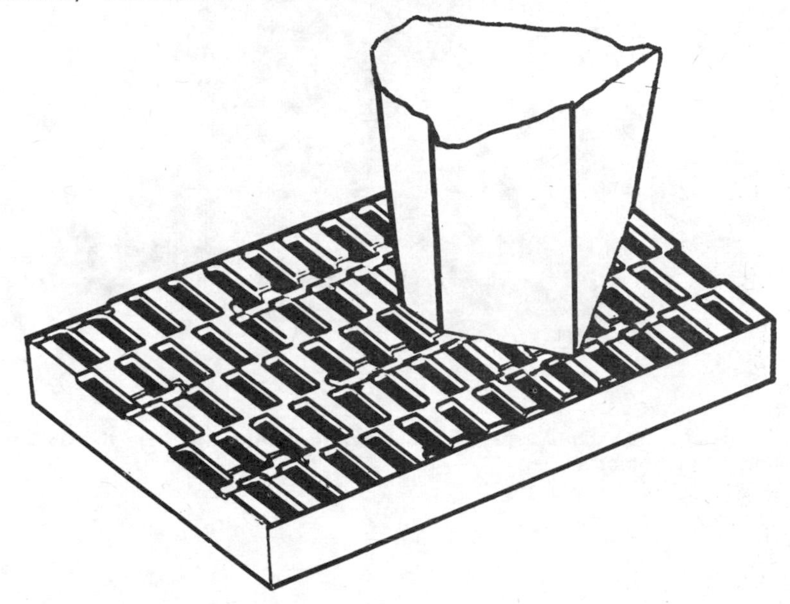

A schematisisation of the VHD, or 'grooveless' capacitance disc developed by JVC.

The VHD disc is encoded less densely than the CED, so it is less vulnerable to wear and damage – although it still travels in a protective plastic sleeve. It is the smallest of the three formats, only 10" in diameter, and the only one which handles both PAL, NTSC and SECAM on one player.

Moreover, the system can handle either high density video (VHD) or audio (AHD) signals. This means that the two types of disc can be used interchangeably on the same player – and that one disc can hold VHD on one side and AHD on the other. With its own AHD adaptor, the disc could even hold computer data, digitised audio or a combination of digitised still frames with compressed audio. (At time of writing, only with such adaptors could sound be tied to a still frame.)

The VHD capacitance videodisc player, with remote control unit, and the disc itself, shown without its protective case.

VHD offers frame accurate random access, two audio tracks, and quick and slow forward and reverse motion. Because, like CED and the CLV laser disc, VHD reads more than one frame per revolution, first generation players cannot satisfactorily freeze on any one frame at random. The picture on the screen will 'judder' (shake or shiver) when the sensor is reading two non-identical frames. This is overcome by recording the same image on two consecutive frames; however, this means that all material to be used as a still or 'freeze' frame must be identified even before the master edit is made. The still frame storage capacity is, obviously, much less than that of CAV laser disc. The system promises to overcome this in later generations.

Both capacitance discs are manufactured along the same lines as the optical discs. The JVC system uses a glass master, the RCA, a metal one, and both use electroform stampers to press the final disc. Both capacitance discs (and Thomson-CSF's reflective optical disc) are compression moulded that is, the disc is made from a 'patty' of electro-conductive plastic pressed flat between two stampers.

The relative virtues of the two capacitance systems, and the laser disc, are discussed from the perspective of interactive video in Chapter 25.

Highlights from the VHD home videodisc catalogue, featuring programmes produced by JVC's European software partners, Thorn EMI.

CHAPTER 16: OTHER VIDEODISC SYSTEMS

As we mentioned in Chapter 13, a great many videodisc systems have been taken as far as the drawing board by various manufacturers and institutions. Some have come very near to the market. Here we are considering two which excited a good deal of interest in their time,

THOMSON-CSF: TRANSMISSIVE OPTICAL DISCS

The French telecommunications company Thomson-CSF developed a transmissive optical disc – that is, one in which light is not bounced off a reflective surface, but passed straight through a translucent disc to a photo sensitive diode on the other side. The two sides of the disc can thus be read consecutively: at the end of the first side, the beam is automatically redirected, and the second side is begun directly without the disc's being turned over.

The transmissive optical disc is manufactured in much the same way as the reflective optical system, and has the same salient features: rapid random access, variable speed and direction, and freeze-frame. However, it does look different, for the transmissive optical disc enters and leaves the player housed in a rigid plastic sleeve, as capacitance discs do.

The Public Archives of Canada used a Thomson CSF player in an early information storage project [1]. Both domestic and 'institutional' players were planned, but Thomson-CSF halted production in 1981. Subsequent agreements with TEAC of Japan, to develop a simpler format, and with 3M, to press discs in the US, were mooted. The company are now making a recordable optical disc, Gigadisc, for the archival market.

McDONNELL DOUGLAS: THE SEE-THROUGH DISC

The American aerospace corporation McDonnell Douglas also developed a transmissive disc, but one which used an incandescent light source instead of a laser beam.

If it ever reaches the market, the disc is intended to be a cheap and handy alternative to existing formats, which can be recorded and replicated in-house quickly and simply on relatively inexpensive equipment. It will handle both digital and analogue signals, and could either stand alone or link to an external computer. Its first appeal is likely to be as a medium for information storage: although it has potential in other fields, early prototypes – like the TeD system a decade ago – could not hold much moving footage.

A handout from the summer of 1982 tells a familiar story: it is scored with handwritten alterations and in many specifications, large, attractive numbers have been struck out and smaller, less impressive ones written in. McDonnell Douglas put the system on ice in the autumn of 1982; however, it surfaced again a year later, still aiming to produce 'desk top copiers' for archival use in business and industry. What happens eventually is anybody's guess.

EPIC's Managing Director, Eric Parsloe, with early prototypes of McDonnell Douglas's transparent disc, and a laser disc from DiscoVision Associates (whose effects are now a part of the Pioneer group).

INTERACTIVITY AND THE VIDEODISC

The Thomson and the McDonnell-Douglas discs are only two of many, but they are enough to suggest the diversity of approaches—and the amount of effort—which has gone into developing videodisc systems, and to keeping abreast of a tough market.

Needless to say, neither of these will be compatible with any other system: the introduction of either would further fragment a market already divided between two types of laser disc and two capacitance disc systems. Whatever else may be said of them, the laser disc, VHD and CED have at least made the long journey from a twinkle in the designer's eye to a prominent display in the video dealer's showroom.

1 Mole, Joseph, and Langham, Josephine. ''Pilot Study of the Application of Video Disc Technology at the Public Archives of Canada'. Ottawa, 1982. (DSS catalogue no. SA2-139/1982)

CHAPTER 17: THE RECORDABLE VIDEODISC

'You can't record a videodisc!' That is a criticism often levelled against discs in the consumer market–and all that many people even know about the disc. In fact, the recordable videodisc is with us now, on the industrial market, and there is every indication that it will be in the home within a few years, as a real rival to videotape. Furthermore, it presages dramatic in computerised information handling.

The videodisc is a cheap, high quality product that resists deterioration and wear, has excellent sound and picture quality, and offers programmes no other medium can produce. In business, industry and education, it is unrivalled as a tool for education, marketing and information storage.

However, the most advanced Level 2 disc players do not yet rival the humblest tape system in one critical respect–they cannot record. In fact, all the videodiscs now on the market are made under extremely stringent conditions in a small number of pressing plants concentrated in America, Japan and north-western Europe.

The manufacturing process is quick and efficient, and not expensive when discs are replicated in large quantities. In a project of any complexity, the time and cost involved in actually making the discs is small compared to that spent designing and executing an interactive video programme. In a straight information storage job, the unit price per disc may be radically less than that of any other storage medium, if a large number of discs are pressed.

However, few information storage jobs run to replication on a grand scale - sometimes a single copy is enough, and a dozen copies are plenty. Often, too, organisations prefer to keep data entirely under internal control. The solution is a system which records and plays back a videodisc as a sort of deluxe microfiche. The first of these was launched in America in the summer of 1983 by Matsushita, under the Panasonic brand name, as the Optical Memory Disc Recorder.

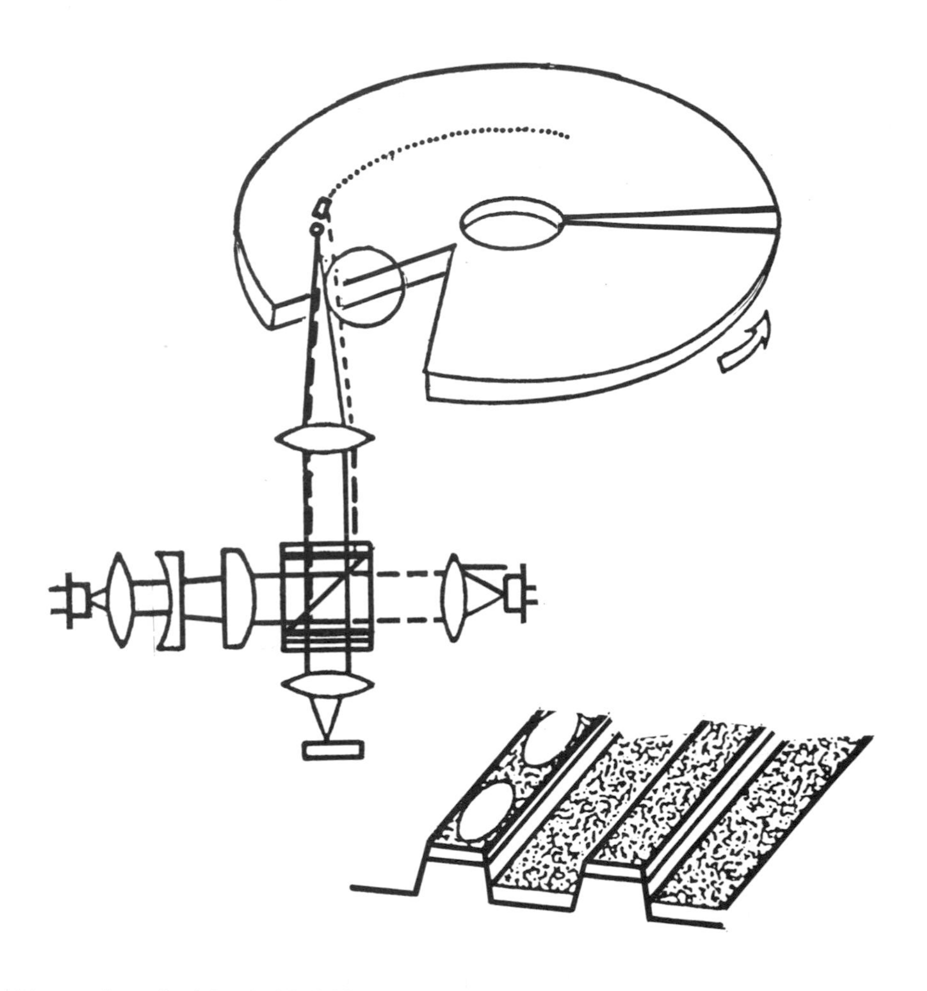

Above, the principles behind Matsushita's recordable disc system. Below, a greatly magnified profile of the disc's surface.

Panasonic's system can record 15,000 frames on one 8″ (20 cm) single-sided laser disc. The hardware comprises a camera, a compact record/playback unit and a monitor, all on the NTSC standard. The camera and monitor are both ordinary models that can be used for other purposes.

There are currently three versions of the record/playback unit. The basic model records and plays back full colour still frames. That designed principally for document storage handles black-and-white still frames only. The third records colour still frames, but can play back motion sequences. Since 15,000 frames moving at a rate of 30 frames a second (the NTSC standard) play through in eight minutes, the emphasis at this point is on archival still frame storage. This, of course, can mean anything from a printed page to a colour slide, a blueprint to an X-ray.

There is a keypad on the player which can be used to call up recorded material, and all the recorders have an RS-232C computer port. This is explained in more detail in Chapter 21; essentially, it is a standard way of linking peripheral hardware to a computer. Thus, the 15,000 frames recorded on the disc can also be organised within a computer-controlled information retrieval system. Typical random access time is half-a-second.

The disc itself is composed similarly to the conventional laser disc described in Chapter 14. It is sealed within a layer of PMMA (polymetha methylacrylate) under a hard polymer topcoat. The spiral track or 'groove' within the disc is stamped on an inner layer of polymer hardened by ultra-violet light during the manufacturing process.

A recording layer of heat-sensitive tellurium suboxide lies within this. The chains of 'dots' which comprise the video signal are recorded in a 'phase change' process. The recording layer starts out in an amorphous state – dull and without definite form. Under intense heat from a semiconductor laser within the recorder, its physical properties change to crystalline – it solidifies, takes on a definite geometric structure, and becomes reflective.

This process is not yet reversible: the recording layer cannot be erased and re-recorded. However, the system employs the DRAW (direct-read-after-write) technique, so errors at least can be nipped in the bud. The system records in 'real time' – that is, there is no need to wait for processing or developing. Immediately a frame is recorded (or 'written', in a term borrowed from computing), it is played back (or 'read') to check for errors. (The laser beam used in the playback process is one-tenth the strength of the recording laser.) The spoiled frame cannot be erased, but its 'address' (the record of its place on the disc) can be, while the data it contained can be recorded again elsewhere on the disc. When the disc is played back, any spoiled frames are ignored.

Toshiba's Document Filing System and Thomson-CSF's Gigadisc are two of the first complete electronic filing systems to be built around recordable optical disc technology.

What we are describing here is analogue information storage, the alternative to which is digital information storage. Analogue and digital are discussed in the next chapter. For our purposes, the first refers to conventional still frame storage and the second to that of digitised information of the kind associated with computer-based technology. This can be virtually anything from numerical data to digitally- encoded video and audio signals.

As we have noted before, the videodisc is only a subset of the larger field of optical disc technology. The optical digital disc can record vast amounts of data, far surpassing the videodisc in density and scope of information storage and versatility. Development of a recordable (and re-recordable) optical disc is a fiercely competitive business. Its immediate object is to substitute optical disc for the magnetic store media (such as tape and floppy diskette) on which computer data storage systems are currently based.

In fact, Panasonic offer an optical digital disc recorder similar to the analogue OMDR system. Furthermore, a large number of companies – including Canon, Hitachi, Shugart, Sansui, Toshiba, Matsushita – are at various stages in the development of microcomputers that incorporate optical disc storage devices.

At time of writing, Canon were actually demonstrating a prototype of such an electronic filing system. While the first stage of development is limited to the storage of black-and-white still frames, such systems will obviously improve rapidly. Sony have even announced a re-recordable disc – that is, one that can be erased and re-recorded.

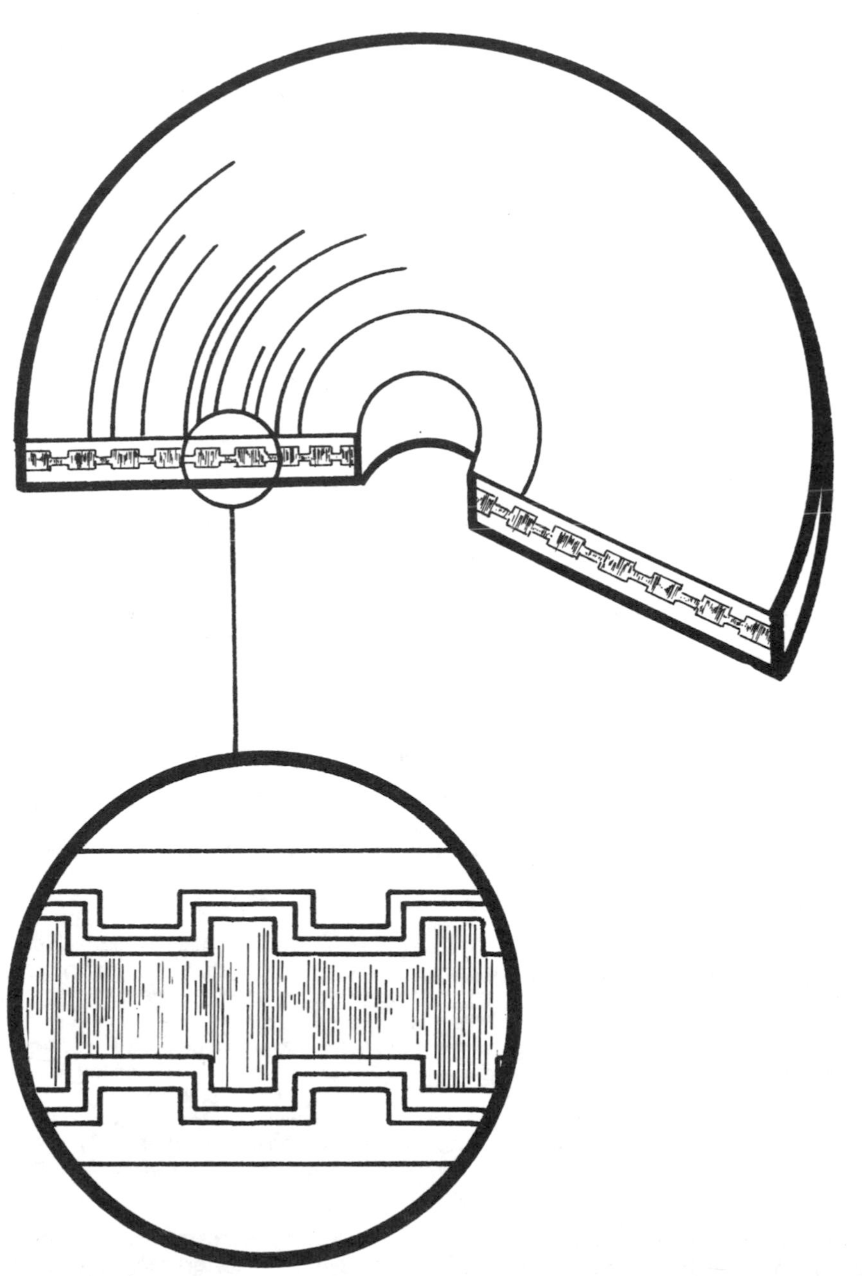

A slice of Sony's re-recordable (that is, erasable) videodisc, in which the recording layer can be formed and reformed more than once.

As we stress in the next section, it is a good idea to think of the videodisc player as a computer peripheral, rather than to regard interactive video simply as an enhancement of conventional audio-video technology. With the development of the kind of technology we are describing here, the disc in fact promises to become the microcomputer's principal storage medium.

Interactive technology now encourages users to put information of one kind (such as pictures and sound) on a videodisc, and information of another kind (such as computer-generated text and graphics, and control programs) on a magnetic computer tape or floppy diskette. With the advent of recordable optical disc technology, it will be possible to store information of all kinds cheaply and compactly, in analogue or digital form, on one medium, and to process that information quickly and easily. There will still be a place for the videodisc, but its more versatile parent promises to be the more important of the two.

The home workstation, so long a popular prediction among futurists (and a feature of our own speculations in Chapter 45), is no longer a distant fantasy, but an objective that could be attained within the next few years. That is, the technology exists: how we will use it is another matter, an open question on which to end this section.

CHAPTER 18: ANALOGUE AND DIGITAL

The contrast is often made between analogue and digital information and technology. The two words appear throughout this book, and in many discussions of new technology. The ideas behind analogue and digital are not easy - information of this kind can only be simplified so far. We appreciate that this may be the most difficult chapter in the book for many readers, and ask only that, if the chapter does give you trouble, that you persevere–read it again, and again. The ideas are basic ones in understanding the new technology–and they are fun once you get the grasp of them. Four basic ideas are presented here: analogue, discrete, binary notation, and digital.

Picture a town in a valley. Imagine that the town's water supply comes from a spring-fed reservoir in the hills. It is important that people in the town know how full the reservoir is, so a monitoring device has been installed. This sends an electrical signal from the reservoir to the town: the strength of the signal indicates how full the reservoir is. In other words, that signal, measured in volts, is the analogue of the volume of water in the reservoir, measured in gallons.

There was once a time when someone just kept an eye on the signal coming down the wire. But now modern technology has stepped in, and the Department of Works has a computer. This keeps a very strict record of water levels in the reservoir over the course of the day, and throughout the year. It does this by taking the level in the first second of every fifth minute and recording this data on magnetic computer tape.

But the water in the reservoir rises and falls constantly, as the spring feeds it and the townspeople draw it off, while the recorded signal is only taken periodically. Thus, each signal represents the water level only as it was at the moment in which the measurement was taken. The record assigns a specific value to a piece of datum taken from a stream of information which is continuously, if sometimes imperceptibly, changing.

The fluctuating level of water in the reservoir is analogue information: it flows, on and on, never the same in two consecutive moments. Periodic readings taken from that flow represent discrete samplings from a long stream of information.

'Discrete' is the conceptual link between analogue and digital, to which we will come in a moment. The two might be compared as a pair of graphs, one with a long, elegant line (analogue), the other as a vertical bar chart with a series of separate, precise values (discrete).

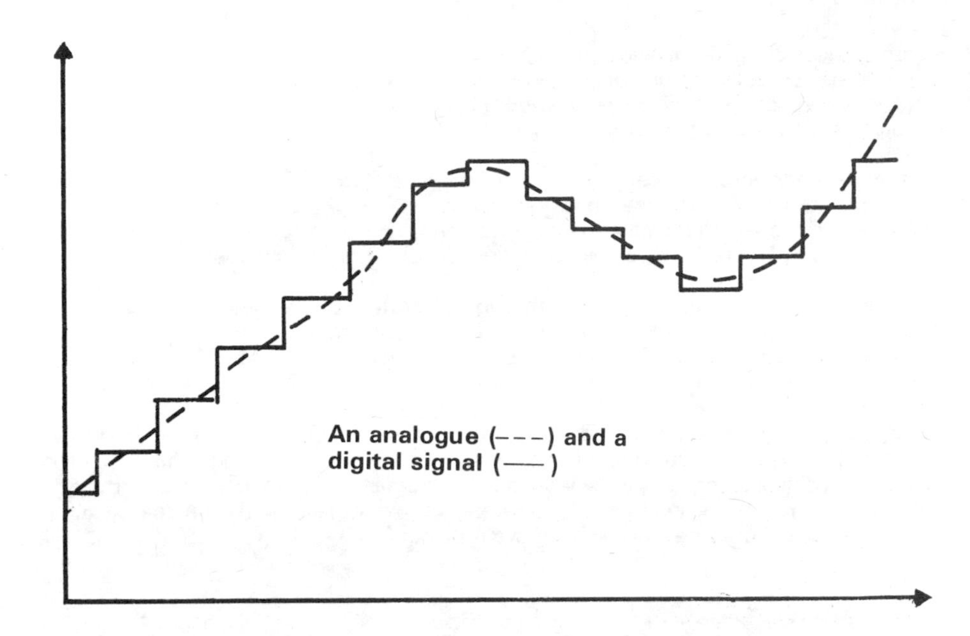

Much information coming from the natural world is analogue: temperature, time, electrical current and voltage – physical properties which are always changing, and which we measure only against an arbitrary scale. Extracts can be taken from such information – a time check, for example, or a temperature reading. This takes a discrete sample from the unwieldy stream, and assigns to it a particular, arbitrary value, such as a numerical code, which is static, stable, and capable of being manipulated.

The word analogue comes from the Greek for 'proportionate', and the word discrete from the Latin for 'separate'. While much of the information we use comes to us in analogue form, the way in which we handle it is in fact discrete. Consider for example various ways of telling the time, and try to deduce which are analogue, and which discrete:

* A conventional timepiece, a wristwatch or a clock, represents the time of day through the relationship between two or three sweep hands, which most people learn to interpret in childhood. Generally, the shortest and slowest measures the hours, the next longest and next fastest, the minutes, and the longest and swiftest, the seconds.

* A sundial (of which there were once many ingenious pocket models) casts a shadow onto a plate which has been marked with radial lines indicating the division of the day into hours.

* An hourglass filters sand through a narrow neck from one chamber to another. A similar device is called a clepsydra – in one type of clepsydra, water or mercury flows from one vessel to another, while in another type, it drips. In each of these, the level of sand, water or mercury in the chambers measures the lapse of time against a pre-determined scale.

With all these devices, we evaluate the position of the moving element in relation to the fixed scale. But only the creeping shadow on the sundial and the clepsydra in which liquid is constantly flowing are examples of 'analogue'. The clock or watch, the hourglass and the clepsydra in which liquid drips, all use discrete units of measurement.

On many timepieces, you can actually see the hands jerking from one position to the next; even those which appear to sweep gently, do in fact move haltingly, dividing the time into fractional but nonetheless discrete units. The grains of sand in the hourglass, and the drops of water or mercury in the clepsydra are, however small, still discrete.

An obvious example of timekeeping through discrete units is the new generation of timepieces which display the time numerically as 'hours : minutes : seconds', so that the time which an aborigine might interpret as 'sun overhead – no shadows', and a passerby might give as 'noon', is displayed as '12:01:13'.

Such devices are usually called 'digital' – not because they use digits alone to display data which might be represented otherwise, but for the technology they use. The essence of digital technology is its dependence on binary notation to handle information of virtually any kind as a pattern of discrete electrical pulses. We have made the jump from analogue to discrete – let us now move onto digital.

BINARY NOTATION

What is intimidating about digital technology is not so much that it converts information into special values, but that it employs binary notation, which can be bewildering to people entirely used to conventional decimal notation.

Binary notation uses only two digits: 0 and 1. When a digital calculator has counted '0, 1', it has used up all its numerals, and from the number two onward has to work out combinations. Decimal notation doesn't have to do this until it reaches the number ten. In binary notation, the sequence '0010, 0011, 0100, 0101, 0110, 0111, 1000, 1001' represents the numbers for which modern decimal notation has the numerals '2, 3, 4, 5, 6, 7, 8, 9'.

The fact that we have ten counting tools at the ends of our two hands probably accounts for the popularity of the number ten as a base for counting systems in so many cultures. But decimal notation is by no means universal, nor absolute (the idea of 'zero', after all, was a relatively late addition to European mathematical theory), nor any less arbitrary than the computer's binary notation.

DIGITAL TECHNOLOGY

The thing about binary notation is that it can be converted into electrical or magnetic signals: 1 and 0, on and off, magnetised and not magnetised. It is basically through 'on/off' signals that digital computers work, and through the 'magnetised/not magnetised' signals that many computer storage devices (such as tape and disk) hold and transmit data. They send a simple message – on/off, 1/0, yes/no. The point is, that they send it with terrific speed, and with considerable accuracy.

The problem with sending an analogue signal through electrical circuits is that it gets distorted. Every time it passes through a piece of electronic circuitry (when, for example, it is recorded, copied, or broadcast), the signal picks up electrical 'noise' that weakens and distorts it. The strength of the signal to the extraneous noise it has accumulated on its way is the basis of the 'signal-to-noise' ratio of technical literature: the higher the number, the better the quality of the signal.

But if the analogue signal is converted to digital code, it can travel a long way without significant distortion: it would take a lot of noise to confuse 'on' with 'off'. Not only numbers, but letters and symbols, audio and video signals, and all kinds of other things can be translated into digital codes - long strings of 0's and 1's. As long as they remain simple electrical pulses, these digital signals do not suffer the steady erosion that analogue ones do. Thus, the digital codes that are reconstructed at the end of the communication chain deliver recorded information as freshly and accurately as if it were brand new.

To return to our example: suppose the reading from the reservoir was '73.5764' - whatever that means. If that were immediately converted into digital code, transmitted as a pattern of pulses, and then converted back into decimal notation, it would still read '73.5764', no matter what distance the signal had travelled, or what processing it had undergone. As an analogue signal, the same message might well come out as '73.58', with a critical loss of accuracy due to the distortion it had picked up all the way.

In our example, we took a reading from the reservoir every five minutes. In practice, discrete readings can be taken far more often than that: the minutely graded vertical bars on an digital chart need hardly differ from the smoothly undulating line on an analogue one – except, if anything, in their greater accuracy.

Digital technology can offer stability and accuracy in many kinds of information handling – including audio and video recording and broadcasting. Specialised data can be handled entirely in digital form – as are the records of things like computer-monitored oil wells.

The detail that goes into digital technology may not concern us, but the potential revolution in information handling may soon do: a world-wide conversion to digital broadcast technology would sweep away many impediments to the rapid, reliable exchange of information across great distances and between cultures. (Ironically, the easier the job becomes from a technical point of view, the stronger may be the political and social resistance to it.)

Standards have already been agreed for domestic televisions which will be able to receive digitally-encoded broadcast signals – only, so far, there isn't much demand for them. Socio-political considerations aside, the advent of digital broadcast technology probably means you will eventually want to buy a new TV - just as you may once have bought a stereo when a hi-fi was no longer good enough.

DIGITAL AND ANALOGUE

Digital technology has brought with it new ways of displaying information familiar to us in other forms. This sets up some interesting problems in perception and interpretation, questions addressed again in the last chapter of this book. As a teaser for that, consider this:

The spread of cheap digital technology is converting the handling of much everyday information from analogue to digital means. We tend to be most aware of this simply in the way information is displayed (17:10 on the digital watch represents the same concept as 'little hand on the five, big hand on the ten' on the grandfather clock, but it's harder to learn to tell the time now than it was ten years ago).

Of course, digital and analogue means of evaluating the same piece of information still both refer to a common scale. If the scale is changed, that, too, is arbitrary, and has nothing to do with technology: 30°F is cold, but 30°C is hot. Adopting digital technology does not alter the information itself, only the handling of it. The work that goes into a digital audio recording is considerable, and the sound reproduction which comes out of it is excellent, but Mozart is still Mozart.

IN CONCLUSION

So, this is the essence of the references you will find to the word 'digital' in connection with many things which do not seem to have much to do with computers. You have just gone in at the deep end. We haven't introduced all the information we might have done (these are supposed to be the gentle, introductory chapters), but having mastered all that, you are now safely launched and ready to handle the really technical stuff. (No, just joking: read on.)

CHAPTER 19: A BRIEF INTRODUCTION TO COMPUTERS

The next few chapters may comprise the simplest thing you will ever read about computers. As we said in the beginning, this is not a book about computers any more than it is a book about video, or instructional design, or screen design or project management. Especially, it isn't a book about computer programming. You need only leaf through a reasonably comprehensive guide to any of those subjects to see why: any one is a career in itself. What this book aims to do is to bring together information about these and other, related disciplines, and explain the basic tenets of each to people whose expertise may lie elsewhere. So, with that qualification, let us consider in simple terms what a computer is, and does.

A popular history of computers is surprisingly entertaining. So is a visit to your nearest science museum. The computer as we know it can trace its family history back to the abacus, if not to the tallystick. Attempts at calculating machines go back at least to the seventeenth century. For a long time, the theory fell foul of available technology. Charles Babbage, the 'Father of Computing', designed his 'Analytical Engine' in the dawn of the Victorian era, but its construction was frustrated by the cumbersome mechanical engineering which was the best on offer in the hoary old days before electronics became an industry.

Computer technology is constantly in a state of flux. Early electronic computers, such as ENIAC (Electronic Numerical Integrator and Calculator), installed in 1946, filled entire rooms. Over the decades, transistors replaced thermionic valves, and were in turn replaced by integrated circuits incorporated onto silicon chips. It is not the technical intricacies of these innovations which concern us here, but the pattern of change which they illustrate.

The rapid growth of computing represents the emergence of an applied science which has profound effects on the workings of our daily lives, but which is still a great mystery to many people. Just as 'engineering literacy' (acquaintance with the basic principles of engineering) became fundamental to general knowledge in the nineteenth century, so 'computer literacy' ought to be general knowledge today.

A dictionary of computing is not only a hefty volume, well beyond the ken of the general reader, but, given the fluidity of language within the computer industry, a work destined always to be incomplete. But there are certain words and ideas used in interactive video technology which it would be useful to discuss here. Some have already been introduced: the last chapter explained the ideas behind 'analogue' and 'digital', and binary notation. Other ideas are explained below; key words are also found in the glossary.

WHAT COMPUTERS DO

A computer, any computer, can be described in terms of four main functions:

* Input: it can receive information.
* Storage: it can hold information.
* Processing: it can manipulate information.
* Output: it can display information.

Input often means typing on a computer keyboard, but may also be effected by speaking into a microphone or drawing on the screen with a 'light pen', by stroking a touch-sensitive screen, or through various input devices such as the joysticks and paddles familiar to computer games. These, and other components of the actual physical computer, are discussed in Section IV.

Storage, or 'memory', most often now employs magnetic tape or disk to record and store information, both data and instructions for the computer. (It is handy to employ different spellings for the magnetic computer disk and the much different audio and video discs, in the same way that it is handy to distinguish computer programs from video programmes.) Paper tape, punched cards, bar codes and optical discs represent other ways, old and new, simple and sophisticated, of holding information outside the computer.

Processing covers a wide, wide range of activities far beyond the straightforward number-crunching many people primarily associate with computers. Literally, everything that happens within a computer is a calculation. However, these functions can be as diverse as editing on a word processor, or sending instructions to a videodisc player. We will discuss processing in more detail a little later.

Output involves the retrieval of information from memory, and the display of that information on the screen and/or its transfer onto some 'hard' medium such as paper.

Of course, computers are increasingly being used to process information like sound, pictures, broadcast signals, and data which are entirely generated and processed in digital form. The structure of computerised information handling is constantly changing. But the uses of the computer in interactive video technology are fairly straightforward, and it is on these that we are concentrating here. Let us consider a few more basic words and ideas.

BITS AND BYTES

We discussed the difference between analogue and digital in Chapter 18, and introduced the concept of binary notation. As we saw there, digital computers ultimately reduce all information to a simple code of zeros and ones, which are transmitted as a pattern of electrical pulses, 'weak' and 'strong' or 'off' and 'on'. These pulses can be sent at great speed without significant distortion. While the way in which a computer handles information may seem positively simple-minded to a human way of thinking, it can work phenomenally quickly by simply shooting electrical pulses back and forth.

One of these pulses is called a 'bit'. This is the smallest unit in computing, and it doesn't tell you much. It takes six or eight bits – one 'byte' – to form a single character such as a letter, a numeral or a symbol like | or &. (Bit is a contraction of 'binary digit', and byte a contraction of 'by eight'.) There is an internationally-recognised code, ASCII (the American Standard Code for Information Interchange), which dictates the combinations of zeros and ones used to form these characters. The number 9, for example, is represented as 111001, and the letter Y as 1011001. This is clear as mud to ordinary human communication patterns, but remarkably efficient in electronic terms.

MAINFRAME, MINI AND MICRO

Computing's heavy-weights are the mainframes – bulky, expensive, and once the only computers there were. They now cater mostly for large organisations needing a lot of computing power and storage, either to control complex projects, or to handle a number of jobs or remote work stations at the same time.

The middle-weights are the minicomputers, which represent the first practical departure in miniaturising computer technology. Minis brought computers within the reach of many organisations which needed computing facilities but which could not justify a mainframe.

The light-weights are the micros, also known as home or personal computers. Micros are also found in a wide range of professional applications, from small businesses with a single machine to large corporations with several different micros doing different jobs in different places.

The bantam-weight microprocessors provide a limited amount of computing power in equipment which is not fundamentally computer-based – certain videodisc players, for example.

The size of a computer is often represented in bits and bytes. This is particularly important since so many new, small computers are at least as powerful as their older, larger predecessors.

* A computer's processing capability – the number of pulses it can handle at one time – is usually reckoned in bits. A typical home computer currently has an 8- or 16-bit microprocessor.

* A computer's storage capacity, or 'memory', is reckoned in bytes or, more properly, in kilobytes ('K'). One K is actually equal to 2^{10}, or 1024, bits. A home computer might currently offer storage of 32K or 64K.

CHIPS WITH EVERYTHING

Electronic computer technology is based on the controlled use of electrons - those particles, found in all atoms, which bear one negative electrical charge. Electrons can be manipulated to produce the electrical signals on which computing is based. Much effort has gone into developing ever smaller electronic circuitry to carry out this electrical information handling.

The heart of the computer is its central processor. When this was composed of thousands of discrete components, valves or tubes, the Central Processing Unit (CPU) was quite large. When transistors replaced valves, it got smaller. Then came the integrated circuit – a complete electrical circuit (that is, the path along which the electric current travels) chemically produced on the surface of a single microscopic chip of silicon. The term 'central processor' became synonymous with the word 'chip'.

Crystalline silicon is, after oxygen, the most abundant element in the earth's crust. In its pure form, it acts as an insulator but, when minute quantities of closely related elements (such as arsenic or germanium) are added, it becomes a semiconductor. That is, it allows the flow of a very small current. The current is measured by the number of electrons that can be persuaded to move when an electrical potential is applied.

The conductivity of a single semiconductor can be controlled by an electrode in contact with the silicon: this is called a 'gate'. Conductivity is high when the gate is 'open', and low when it is 'closed'. The open and closed states are obtained by applying a high or low voltage. The application of binary logic to this state of affairs is obvious.

Each of these gates controls only one bit – high/low, 1/0, on/off – but many, many bits can be packed into a very small space, very cheaply. Hundreds of thousands of sub-microscopic ICs (integrated circuits) fit onto a single chip. In fact, before the end of this century, the number of transistors on one chip is expected[1] to equal the number of neurons in the human brain. It is easy to see why the use of chips goes beyond the computer industry into virtually anything requiring electronic control circuitry, from hardware for outer space to pens that tell the time.

PERIPHERALS, INTERFACE AND SOFTWARE

Of the computer's four basic activities, processing alone takes place entirely within the computer proper, in its central processor. But the central processor isn't much use on its own: input, storage and output all depend on other hardware – the computer's 'peripherals'. Typical peripherals include the computer's screen, keyboard and printer. But every object that communicates with the computer (including the video player and monitor of an interactive video delivery system) must link to the central processor, and can be described as peripheral to it.

This link is accomplished through an 'interface', an all-purpose word that describes both the principles and the combination of chips, wires, sockets, cables and plugs which actually carry information between the central processor and its peripherals. Interfacing, and how to effect it, is discussed in more detail in Chapter 21.

Of course, all this is still hardware, inanimate and unintelligent: what makes a computer go is the complement of hardware, software. Software describes the programs which actually control the computer as a whole, as well as those which do specific jobs such as running an interactive video programme.

We will look at software more closely in later chapters. First, let us consider how all this wizardry works in practice.

CHAPTER 20: HOW COMPUTERS WORK

What we have explained up until now has been simplified. What follows now is breathtakingly simplified.

A systems analyst will find this section wanting, just as a studio technician will find the section on video a little thin. But were those two brought together on an interactive video project, the systems analyst might learn something from the chapter on video, and the studio technician might pick up some information here. Which is all we are trying to do. This chapter is not a handbook on building computers for fun and profit, but a cursory examination of how computers process information, and what stands between the person in front of the screen and the chip in the heart of the system.

When all computers were mainframes and ordinary people had little to do with them, computers were dubbed 'superbrains' and there was a lot of speculation, humorous and otherwise, about their ability to conquer the universe. Now, all kinds of people with no formal computing background are casual users, mostly of microcomputers, and it's an accepted truism that computers are 'dumb'–'garbage in, garbage out'.

Computers interpret information in ways which are quite different to the workings of people's minds. Analogue computers convert information into electrical signals, variables such as currents and voltages. Digital computers assign precise numeric codes to information which we perceive variously as textual, numeric, aural or visual. Not surprising, then, that people communicate with computers in unique and special ways, too.

There are, moreover, different levels of communication between people and computers. We are not introducing these things to frighten those readers who are not ace computer designers, but merely to lay the groundwork for what follows. Keep calm.

MACHINE CODE

Each machine has its own 'machine code' (or 'instruction code', or 'microcode'), with which it works in the central processor. This is a numeric code unintelligible to any but its designers, and hardly a practical medium with which to work from day to day.

To make life easier, software designers have developed ways of writing machine code so that it looks less intimidating. One way of doing this is to take an instruction written in machine code as a group (or groups) of eight bits, and to represent it in familiar letters. These might spell out short words, or abbreviate long ones.

For example, in one particular chip (the Zilog Z80 is its name), the machine code instruction to add two numbers together is '11101101'. This is represented as the word 'ADD'–which, you must admit, tells you a lot more than the instruction to '11101101' a pair of figures.

There are many more instructions of this kind–'JMP' for 'Jump' or 'CP' for 'Compare', for example. The whole lot is called an 'assembly' language. A special piece of software called–wait for it–an 'assembler' translates these alphabetic commands, which are clear to a human mind, back into the machine code which is clear to the computer.

HIGH LEVEL LANGUAGES, COMPILERS AND INTERPRETERS

Only an adolescent 'superbrain' who makes £1,000,000,000 a week designing computer games would actually write an assembler for fun. Certainly no one else would understand the programs to result from such an exercise. To cater for the bulk of us mere mortals, the same software designers who wrote the assembler have also devised a method for communicating with the computer using everyday words and symbols.

You'll have to take our word for it, but the sequence which in the assembler looks like:

```
LD      A,&H42

ADD     A‡5

LD      (αH44),A
```

. . . is just about the same thing as a statement like:

```
LET     X=Y+5
```

. . . in a high level computer programming language like BASIC (Beginner's All-Purpose Symbolic Instruction Code). Paradoxically, a 'high level' language is one which looks quite simple to the likes of us, while a 'low level' one looks terribly complicated. That is because the perspective is not that of the end user, but of the software designer who has to create the language. The nearer a computer programming language is to ordinary human speech and logic, the farther it is from machine code. As is often the case, the simpler it looks, the harder it was to do.

Of course, for the computer to understand these commands, every statement in a BASIC (or other computer language) program is translated into machine code, again by a special software program, this time called, appropriately, an 'interpreter'. An interpreter translates each line of, say, BASIC, and executes commands (such as adding up a column of figures or displaying a calculation on the screen) when instructed to do so. One line of BASIC is typically equivalent to three to seven lines of assembler.

In some languages, the whole program is translated before any action is taken. This process is called 'compilation', and is accomplished through a special program called a 'compiler'. Compiled languages tend to run faster than interpreted ones, and are a feature of larger computers.

OPERATING SYSTEMS

Each machine also has an 'operating system', which controls the way in which the machine works, and defines how it and its peripherals can be used. Each machine could have its own unique operating system, and many do. If this is the case, then all software for that computer must be purpose-built to communicate with its unique operating system.

Designing computer programs from scratch, especially for basic things like book-keeping or word processing, is not a practical activity. So many microcomputers employ a standard system – CP/M (Control Program for Microprocessors) is currently a popular one. This means that they can use a wide range of commercially-produced software packages which are much cheaper and easier to implement than purpose-built ones. Of course, special programs may still be designed for special jobs. Some computers use both a unique operating system for their own private software, and CP/M.

The operating system defines the ways in which the chip at the heart of the computer can interact with the outside world. It dictates how the computer's peripherals – the screen, the printer, the videodisc player and so on – can be used, and what applications programs are compatible with it. It also executes a number of 'invisible' functions which are a vital part of day-to-day operations.

For example, the operating system constantly monitors the system and its operations. It notes where a storage device such as a magnetic computer disk may be worn or damaged, and ensures that information is no longer stored there. The more checks that the operating system carries out, the less likely the system is to 'crash', or break down, through mechanical failure. Even when there is a problem, the operating system can often detect the fault, and transmit a message through the screen or the printer which will help the operator or engineer effect a recovery.

The operating system is, therefore, an important part of the computer system as a whole. From the point of view of an interactive video application, the main point is simply to ensure that the software which is used in a Level 3 system is compatible with the computer's operating system. Since most interactive video delivery systems employ well-known micros, and software designed to appeal to as wide a market as possible, this is rarely a large problem.

COMPUTER PROGRAMS

Then there are the computer programs themselves. These comprise long, detailed strings of instructions which take the computer through a series of activities which, to the human mind, often seem painfully convoluted. These are, in fact, often so relentlessly linear that they seem illogical to minds accustomed to less consistent ways of thinking.

For example, a person looking for a word in a dictionary will move directly to the section comprising words with the same initial letter as the word sought, then to those with the same second letter, and so on, until the very word has been found. But a computer might start at 'Aachen' or 'aalii' or whatever, and examine every single word in turn until it finds the one that matches exactly the one it was asked to trace. This would not be a practical way for a person to work; however, a decent computer could go from 'aardwolf' to 'zymurgy' in the time it would take most of us to pick up the dictionary. The way in which computers handle information is suited to the way in which they work, and to their essential lack of creative intelligence.

An actual program, for instance, might look like this:

```
 90   N=1
100   OPEN ''MYFILE''
110   INPUT ''Name to search for?'';N$
120   READ MYFILE,M$
130   IF N$=M$ THEN GO TO 150
140   N=N+1 : GO TO 120
150   PRINT ''Name was at position'';N;''in the file''
```

Computer experts would agree that this is not a very good program, but it is sufficient to show the untrained eye what a typical sequence might look like.

But, take heart—there is a short cut through all this. It's called authoring. Computer programming—and that alternative to it—are discussed again in Chapter 22.

DEDICATED AND UNIVERSAL

There are generic programs, such as database packages or accounting packages, which handle routine jobs in a way acceptable to a wide range of users. Software distributors retail these in much the same way that bookdealers handle books. Then there are 'dedicated' programs, designed specifically for one particular job on one particular type of computer. These are generally expensive and time-consuming to produce, for computer programming is a labour-intensive activity and computer programmers are highly trained people. This is why many users are happy to use commercially-produced generic software—often called 'applications programs'—on routine jobs.

In the same way, there are both 'dedicated' and 'universal' computer systems. An airplane's autopilot, for instance, or the processor in a purpose-built interactive video delivery system, are designed to do one job and one job only—dedicated to it, in fact.

However, in business as well as in the home, one computer is often expected to handle many different jobs, sometimes juggling several at one time. A home computer usually employs different software for different jobs. A package designed to produce computer graphics or to do word processing comes with a computer disk which tells the computer what it needs to know to do that particular job. This is inserted in the machine to set it up for the job it is about to do. The actual work – the graphics or the word processing – is usually prepared on a separate, 'working' disk, which is simply a storage device.

On a larger computer, users on a whole network of remote terminals may draw on a single central computer. It is not that the computer can do two things at once, but that it does each job so quickly that it can keep up with a number of different users doing different things, by giving them each slots of time. What ever 'time sharing' may mean on the Costa del Sol, this is what it represents in computing.

This, then, is a grounding in some of the ideas which are germane to a basic understanding of computer technology. Some of these points, as we noted, simply prepare the ground for the information and the arguments which are to follow. Let us now consider subjects which are a vital part of interactive video technology – computer/video interface, authoring systems, and computer graphics.

CHAPTER 21: VIDEO/COMPUTER INTERFACE

We are now discussing projects on Level 3 of the scale described in Chapter 5: video players and screens under the control of an external computer.

'Interface' is a word you might find in chemistry, where it describes the surface that separates two phases, or in a dressmaking, where it means the stiffening fabric in lapels and cuffs. But its best known application is to computers:

* Specifically, interface means the connection between the computer's central processor (the chip at the heart of the system) and its peripheral units - that is, the rest of the computer: its screen, keyboard, disk drives and so on.

* Generally, interface describes the connection between any two pieces of equipment which have to communicate with one another.

Like a number of words in the vocabulary of computer technology, 'interface' has wormed its way into several parts of speech. It is used as a verb and as a noun, and an adjective, sometimes, too.

Interface is the central processor chip's link to the outside world. In this sense, it is an integral part of the computer system, developed along with other aspects of its design and construction and incorporated into the package when the computer is put on the market. Interface at the basic level is an 'invisible' component of the computer. It costs a good deal of time and money to develop, but once it is perfected, an interface between the principal parts of the computer is something about which the average user never need worry.

However, an interface between the computer and new or unusual peripherals—such as a video tape or disc system, or the hardware in a simulator—is not always so straightforward. In the early days of interactive video, the development of computer/video interfaces was inevitable in many pilot projects, and contributed significantly to the length of the schedule and the cost of the programme.

Fortunately, the way has been smoothed in the last few years by the introduction of any number of commercially-produced interface packages, which link most of the popular computer and video systems routinely, at a modest price. Someone who enjoys electronics as a hobby could make a computer/video interface package at home for fun.[1] However, most users are more anxious to get on with making a programme than to play with the hardware, so we will concentrate on 'off-the- shelf' interface here.

SYNCHRONOUS AND ASYNCHRONOUS COMMUNICATIONS

Interactive video is very much concerned with way in which the computer communicates with its peripherals. Timing is critical when we consider that a single second of video contains over two dozen separate frames (25 in PAL, 30 in NTSC). A split second can mean the difference between a clean cut to the first frame in a scene, and a sloppy one which starts on the last frames of the preceding segment.

Ideally, the computer and the video player should 'talk' to one another - communication should be 'bi-directional'. Typically, the computer issues an instruction to the video player, such as 'Play from frame number 32,156 to frame number 36,829'. The video player then executes that instruction: it plays the segment, starting and stopping on the very frames it was instructed to identify. Immediately it has done this, it 'tells' the computer that it has finished, and is ready for another instruction.

This is 'asynchronous' communication. The completion of one operation initiates the next, no matter how long each one takes. Since computer programs must spell out every discrete step in painstaking detail, several instructions in rapid succession might be needed to set up a job which will then take some minutes to execute. So long as the computer and its peripherals are in free communication, the sequence and accuracy of these operations are guaranteed.

In 'synchronous' communications, there is effectively no two-way dialogue between the computer and the video player. ('Synchronous' derives from the Greek 'with time'.) The peripheral unit cannot tell the computer when it has completed its assigned task, and is ready for another. So the computer has to send successive instructions at arbitrary intervals, more or less accurate. The program designers estimate how long each operation should take, and when the peripheral is likely to be ready for another instruction. This adds work to the preparation of the computer program, and the end result is less accurate and decorous than it would be with asynchronous working. This also represents one real difference between domestic and industrial standard video equipment: with many domestic players, only synchronous communications are possible.

FRAME ADDRESSING

A key phrase to look for is 'frame accuracy' or 'absolute frame addressing'. This means that any individual frame can be identified and found accurately every time it is requested. Without this feature, the desired frame may be found more or less or exactly, but cannot be guaranteed.

Each frame in the video (and each byte or word of computer data) has its own 'address'–the slot assigned to it on the storage medium. This is rather like a post office box number, which is an integral part of the structure of the storage medium, and stays the same no matter what information is put into the box or slot.

INTERFACE BOARDS

The object which makes interface possible is called a 'board' or 'card'. It is simply that–a flat board about the size of an outstretched hand. On it are several small rectangular plugs, looking rather like robotic millipedes, with wire 'lugs' along both sides like tiny stilts or legs. These hold the chips which control the computer's operations and memory. There are also various transistors and fuses, and a number of channels along which connecting wires run, so that the whole thing, laid flat, often looks like an aerial photograph of an industrial estate, with warehouses, storage tanks and roads.

A separate set of lugs connects the board to the computer, so chips with complementary functions on different boards can communicate with one another. Various cables, fat and thin, round and flat, like so many varieties of pasta, run off the board and out the back of the computer to plug into various peripheral units in the same way that various wires connect the discrete components of a home stereo.

Most microcomputers have room for a limited number of these boards (currently, four to six is typical). Some chips address vital functions, and must always be present in the machine. However, others fulfil special functions–like driving a video tape or disc player–which some users will want and many won't. There is usually room for at least a couple of these boards, which can be taken in and out routinely. You can see the cables emerging from these boards through the slots on the back of a home computer. Inside the machine, the boards themselves are lined up like the garden walls of so many suburban houses.

COMPUTER PORTS

So, communication between the computer and its peripherals travels along cables very like the ordinary electrical cables that connect the equipment to a power source such as a wall socket. The connection between the computer and its peripherals is usually effected at a special socket called a 'computer port'. There are different kinds of computer ports just as there are different kinds of operating systems, some purpose-built and others standard.

Standard ports make computer terminals and peripherals as easily interchangeable as, say, the disparate components of a stereo system. With standard interfaces, computer owners have a choice of printers just as hi-fi enthusiasts have a choice of speakers. Instead of being tied to one manufacturer, users can shop around for the peripherals best suited to their needs and budgets, and there is room in the market for small companies as well as large ones.

Through standard interface, new peripherals such as video players can easily be integrated into existing systems. This considerably enhances their market potential, for it means that an interactive video delivery system can usually be cobbled together from computer and video equipment already at hand, for little more than the cost of the interface package. This is coming down all the time, for there is fierce competition both to reduce prices and to improve the product. The $100 interface package is currently the rock bottom goal. Once one of the thorniest parts of a Level 3 application, video/computer interface is now fairly routine and relatively inexpensive.

The idea that video equipment is just another computer peripheral is a useful attitude to take not only toward engineering, but also toward the role of video in interactive technology. Many people approach interactive video from as experienced video users and producers, only to realise that it is essentially a computer-based discipline that treats video—the be-all and end-all of modern communications technology—as just another tool. Looking at interactivity as a new category, with its own aesthetic and standards, is a much better perspective than treating it as the very latest in magic lantern shows.

SERIAL AND PARALLEL

There are two main ways of sending communications between disparate pieces of hardware. A 'serial' interface sends information down the wire one bit at a time, with all the bits in a byte or word travelling in single file. A 'parallel' interface sends data one byte at a time down a group of parallel wires, so all the component bits in a byte travel abreast. Different makes and models of industrial standard videodisc players, for instance, feature ports of different types. For this reason, some commercial interface packages are compatible with ports of several kinds.

The most popular standard port in computers of all sizes is currently the 'RS-232C' serial interface. Generally, any two devices with RS-232C ports simply plug together without any interjacent hardware. Most of the new generation of personal computers, and some industrial standard videodisc players, feature an RS-232C port. Interfacing between them is a matter of cables and plugs. However, older models and domestic standard video disc and tape players are not so accommodating, and an intermediary device is needed. This link is commonly an object known as a 'black box.'

THE BLACK BOX

The black box stands between the computer and the video equipment, and 'translates' between them. 'Black box' is a generic term, like 'box of tricks', defined by the rather trendy Collins English Dictionary as ''a self-contained unit in an electronic or computer system whose circuitry need not be known to understand its function.''

The box in Allen Communications' VMI (Video Microcomputer Interface) really is black, and fits in the palm of the hand. The box can, of course, be any colour - and is usually much larger, particularly in those packages which are 'controllers' rather than 'interfaces'. Controllers, such as CAVRI's Intermedia Universal Video Controller, integrate both the interface (hardware) and an authoring system (software) in one package.

VIDEO/COMPUTER INTERFACE PACKAGES

It is, then, quite possible and entirely advisable to buy an interface package off the shelf. There are a number available, and more emerging all the time. As we said, while a small part of the computer market as a whole, this is a field of rapid growth and strong competition. Development is concentrated in America, which is still a focus for interactive video technology as a whole.

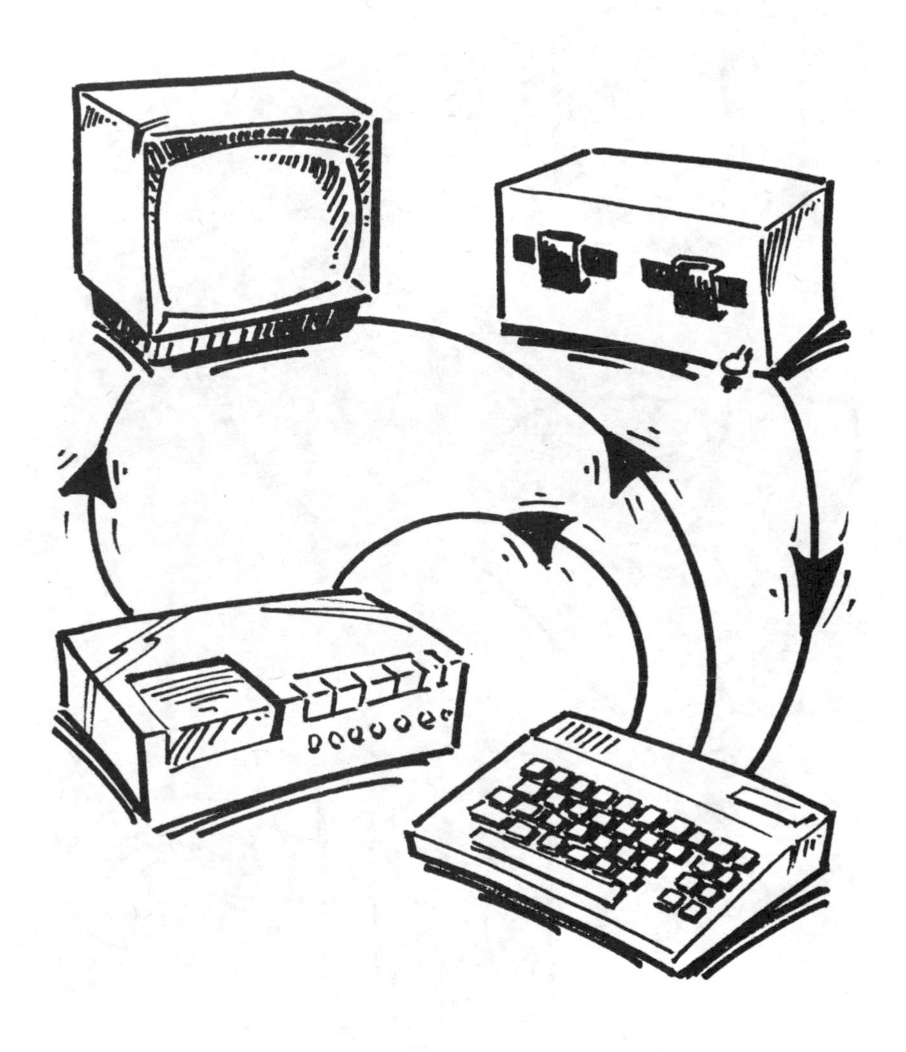

A typical tape-based Level 3 configuration.

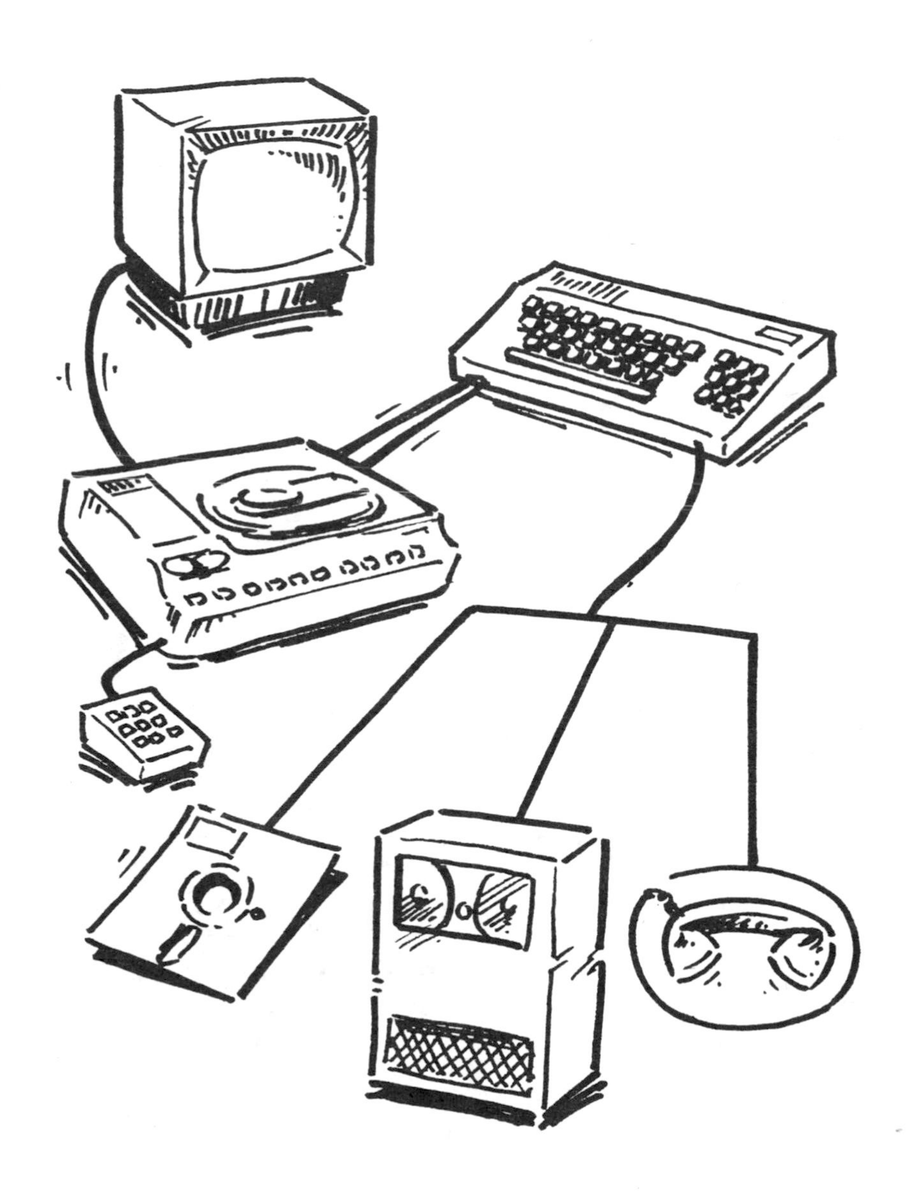

A possible disc-based Level 3 configuration.

Some packages offer straightforward interface, while others are virtual delivery systems, including not only hardware to link the video and computer equipment (and sometimes, even the equipment itself), but also the software neeed to make interactive video programmes. Those presented as complete delivery systems are discussed in Section V. However, as many of the companies who first introduced basic interface packages are now offering complementary software, the distinction is no longer clearcut, so we will cite 'controllers' as well as plain interface packages here.

We have mentioned Allen Communications, who are well established in this market:

* The initial VMI package links the Apple II microcomputer with NTSC laser disc players.

* The subsequent Universal Video Controller links virtually any computer (mainframe, mini or micro) with a number of specified video disc and tape players. It can handle more than one player at a time, and, as a controller rather than a simple interface, allows processing of video information through the computer.

On the other hand, some packages are based on two specific pieces of hardware:

* The Discmaster series, by New Media Graphics, is built on an Atari computer and a Pioneer NTSC disc player.

* Omniscan, by Aurora Systems, links the same Pioneer player to an Apple II.

Some companies offer packages built around their own equipment.

* Philips' 'professional' videodisc system even offers an optional CPU right onboard the disc player.

* Pioneer have their own black box, developed in Japan and based on their laser disc equipment.

* Sony, who make both an NTSC laser disc player and a microcomputer, offer the two as a complete delivery system. Sony also market The Responder, a delivery system based on their U-matic videotape recorder.

* National Panasonic also offer a system which can use their tape equipment or that of other manufacturers.

* Computer manufacturers like Wicat, Positron and Tandy work the other way and supply interfaces based on their computers.

Lately, the trend has been toward versatile interface systems based on the RS-232C interface.

* CAVRI's Universal Video Controller, mentioned earlier, is compatible with any computer that has an RS-232C port, and a wide range of videotape players.

* BCD Associates first introduced an interface to tie video tape machines with the Apple II, and then went on to produce Video Link 232, based on computers with RS-232C ports and videotape players.

This is by no means a comprehensive list, but does cite some of the better-known systems on the market at the time of writing. Again, consultants, technical magazines and the manufacturers themselves are your best source of up-to-date information.

AUDIOTAPE AND VIDEODISC

Another approach altogether, and one which promises to bring interactive technology within the reach of a wide market, is based on audio tape and Level 2 industrial standard videodisc players. These players can be programmed through the remote control keypad, but such a program cannot easily be edited or amended, and is lost every time the power is switched off. So while fairly complex programs can be designed through the keypad, the limitations on editing, and the ephemeral nature of the undertaking, tend to frustrate its being used at its full potential.

An interface developed by the Centre for Aerospace Education at Drew University offers a low-price solution. The interface connects to the disc player, its remote control keypad and an audio tape cassette recorder. It intercepts programming instructions entered through the keypad, and records them. Thus, any number of programs can be stored at one time, and easily retrieved and even edited.

Since it can also record straightforward audio commentary, the system has been dubbed 'the $200 solution to sound-over-stills'. Even with the radically condensed audio signals of 'compressed audio' systems, it is not yet possible to fit more than a few seconds' audio onto a single frame of video when the two are recorded together. The difficulty of incorporating a complementary audio commentary does hinder the use of disc's facility for still frame storage. This system, in which the audio commentary and the videodisc control program are recorded together, represents a valuable departure on conventional tape/slide presentations as well as on other applications of interactive video technology.

INTERFACE AND DELIVERY SYSTEMS

Pre-packaged delivery systems, that come complete with all the requisite hardware and software, include interfaces between their component parts. So do products which are essentially controllers in fancy dress, such as the Sony Responder. These are discussed in Section V.

Then there are the various reprocessing systems, which allow signals from both the video player and the computer to be displayed on one screen. These are discussed in Chapter 20, along with computer graphics generally.

The interface packages cited here address a wide selection of computers and video tape and disc players. However, it should be clear that this is a changing market, and that many systems, although versatile, do have limitations. While there has been a concerted move toward standardisation by both video and computer manufacturers, there are still quirks in the design of many machines. The only way to be absolutely sure that two pieces of hardware will work together is to consult the dealer or manufacturer directly, citing exactly what makes and models are involved, and what is expected of the final delivery system. To that end, addresses and telephone numbers of some of the makers and distributors of video equipment, delivery systems and interface packages are listed in the back of this book.

1 Ciarcia, Steven A. 'Build an Interactive Videodisc Controller'. Byte, June 1982, page 60.

CHAPTER 22: AUTHORING SYSTEMS

First, a word about computer programs.

A computer is a tool, an inanimate lump of hardware. What makes it go is software—particularly, computer programs. These comprise long, detailed strings of instructions which take the computer through a series of activities which, to the human mind, often appear painfully detailed. This is because computers are relentlessly linear, and lack what we consider to be creative intelligence. A computer program will spell out every discrete step in an operation which a person would think of as a single coherent action. (Think about any familiar physical or intellectual activity from this perspective and the simplest task becomes monumental.)

That computers have their own languages, logic and notation is unavoidable because of the way they work, but it does tend to cut them off from people lacking the requisite communications skills. Computers don't 'think'—they are, after all, machines operating through programs designed and run by human beings—but the way in which they process data is unlike the ways which people often take for granted as being the 'only' approach to information.

Data and programs are usually input through some familiar medium of human information exchange such as the alphabetic codes of the written word or the numeric ones of arithmetic. As we saw in previous chapters, this information is converted within the computer into the code of electrical signals which is the computer's own medium of communication. The computer can then store and process this information in a form which it understands. However, to be of use subsequently, the information must be converted back into a form which people can understand, whether words, symbols, sound or pictures.

Computer programs are written in computer languages by computer programmers. Popular languages in microcomputing are BASIC (Beginners' All-purpose Symbolic Instruction Code) and Pascal (which recalls the seventeenth century French philosopher, mathematician and physicist who built the first mechanical calculator). Specialist languages include COBOL (Common Business Oriented Language) and the scientific and mathematical languages ALGOL (Algorithmic Language) and FORTRAN (Formula Translation). Although these languages tend, like Sanskrit, to be written rather than spoken, they still have to be learned, and fluency in one or more is an attribute of a qualified programmer.

A computer program is not immutable: it can always be updated or amended. However, programs of any complexity are difficult to put together, and even more difficult to change subsequently—especially by someone other than the original programmer. Computers may be rigid and logical, but computer programs tend to be highly individual.

People make them according to their own needs and perceptions, and a computer program is often a fairly good reflection of how the mind of the computer programmer works. Computer programming can be creative and satisfying for someone designing a program, for the same reason that it can be intensely frustrating for the person trying to alter one.

This is why people tend not to qualify as computer programmers without some substantial reason for doing so. Which, in turn, is why—in education and training particularly—people with expertise in their fields often still depend on programmers to prepare courseware for computer-based training systems. This is also why the computer industry, growing ever more egalitarian in response to an emerging market of casual and domestic computer-users, saw the wisdom in offering an alternative to full-fledged computer programming. This is 'authoring'.

AUTHORING

Authoring represents a shortcut through computer programming. The people who use 'authoring systems'—'authors' or programme designers—do not have to qualify as computer programmers.

Some authoring systems have their own 'authoring languages', which, although nothing more or less than computer languages, are simpler to learn and to use. Some authoring systems are even presented in everyday language (for the purposes of this readership, English).

Authoring systems employ 'editors'. These are programs which encapsulate in one straightforward command from the user to the system, the long string of instructions to the computer which is inevitable in a computer program. Of course, operations are still carried out in the same way within the computer—the difference is that the author is not bothered with the details. The authoring system makes that leap between computer logic to human logic. The author has only to think about activities which directly affect the production of the finished programme, and not those which concern only the computer's central processor.

Authoring systems tend to employ terminology which reflects their wide use in the preparation of 'courseware' in computer-based training and education programmes. However, programmes of many kinds can be designed through authoring systems. The general term 'information programme' describes the wide range of applications concerned with 'information transfer'—catalogues, point-of-sale units, and so forth.

Authoring languages are themselves 'very high level' computer programs—that is, they are far removed from the 'machine code' at the heart of the computer. Authoring programs are designed by computer programmers and written in computer languages like BASIC or Pascal. Usually, an 'authoring system' is a set of complementary programs which provide the programme designer with a kind of kit from which courseware or information programmes can be constructed. Because they are ultimately written in a computer language, authoring systems can be prepared or customised by qualified programmers to address special needs or applications.

An authoring system often comes as a set of floppy diskettes together with an instruction manual. These diskettes are used like those used in a financial package or word processing package or any other generic software package – they tell the computer how to do the job asked of it.

Typically, an authoring system comprises a 'preparation system' and a 'delivery system'. The preparation system is used to design courseware – that is, the program which is ultimately presented to the end user, whether student, employee or consumer. The delivery system is used to run that program on a day-to-day basis. This delivery system is often built into the individual 'lesson diskette' on which the program is recorded and with which it is run.

Authoring systems are sometimes sold outright, and sometimes 'licensed'. Either way, their designers are anxious to recoup their development costs. The cost of an authoring system must be seen in relation to the cost of designing a computer program – or several – for all the courses or information programmes which can be created through a single authoring system.

SCREEN-BASED AND INSTRUCTION-BASED SYSTEMS

Authoring systems are currently basically of two types – those which are used to design a series of individual 'screens' on the computer's VDU, and those which build up a program as a long string of instructions in the manner of a simplified computer program. These are referred to as 'frame-oriented' and 'line-oriented' systems, respectively.

'Line oriented' systems generally have their own languages, similar to or based on computer languages. Apple's SuperPILOT, for example, uses a language close to BASIC.

In a 'frame oriented' system, authors actually work on the screens which will appear in the finished programme. These screens may hold text, questions, instructions, graphics or any other computer-based information. The relationship between them, outlined in a separate stage of the authoring process, dictates the branching patterns and indicates when segments of video will appear between screens of computer- generated material. Typical screen-based systems are Microtext, developed in the UK by the National Physical Laboratory, and The Instructor, developed in America by BCD Associates, and taken up in the UK by CP Software as IVL (Interactive Video Learning).

The two approaches simply represent different perspectives on the same activity. Line-oriented systems employ long lists of commands to the computer to design both the structure of the programme and the appearance of individual screens. (The author can, of course, look at these screens – to see how they are developing, and perhaps then to modify the instructions which describe them.) Frame-oriented systems are built directly around the screens themselves, and display only those instructions which relate directly to decisions made by the designer – those which dictate the order of segments, for instance, or branching patterns. Of course, frame-oriented systems ultimately depend on long strings of instructions to the computer just as line- oriented systems do – the difference is that these instructions are not displayed in a frame-oriented system, and the author generally neither sees them nor directly refers to them in the design process.

Line-oriented systems generally take longer to learn and to master than screen-based systems, but are usually concomitantly more versatile. Some systems, such as McGraw- Hill's Edutronics and Bell and Howell's AVA (Audio Visual Author), offer both approaches. AVA, in fact, allows the author to jump from the instruction file to the screen using a single key on the keyboard.

USING AN AUTHORING SYSTEM

Authoring systems are often 'menu-driven'. That is, the author works with a screen that lists the options from which the programme can be designed – rather as a restaurant menu offers the diner a choice of dishes from which a meal can be ordered. The menu of an authoring system is usually shorter than that of a restaurant, although the 'meal' composed from it can be a veritable banquet.

```
CAVIS EDITOR    EDIT OPTIONS

1  Look at page
2  Create  page
3  Alter   page
4  Copy    page
5  Load    presentation
6  List    presentation
7  Create  presentation
8  Alter   presentation
9  Save    presentation
10 Run     presentation
11 Prepare new cassette
12 Reset system disk (drive 1)
13 Reset presentation disk (drive 2)
14 Monitor external Video input
15 Receive page from external equipment
16 Send page to external equipment

SELECT OPTION :
```

A typical menu screen from the CAVIS system.

The options on the menu represents a set of templates for the elements from which the course or information programme can be constructed. Programmes designed with an authoring system tend to be modular, built around 'information blocks' which are in turn composed of any number of individual components. In an interactive video programme, these could include still and moving video footage and/or computer generated text and graphics. Information blocks exclude menus, tests and so on–elements of the programme that do not directly convey information.

A programme executed through an authoring system has to be as carefully thought out as any other. The authoring system is simply a tool which facilitates the construction of the interactive programme. It executes the work of the design and flowcharting stages described in Section VI. Although it is possible to make changes once the programme is set up, this is a wasteful and frustrating activity when the fault lies in poor initial design.

In a frame-oriented system, the first stage is usually to outline the shape of the course, 'filling in the blanks' on screens which outline the basic conditions and structure of the system. Each block must be named or numbered, and every segment in every block similarly named or numbered. The running order of the programme and the branching pattern are established at this stage. These details can be altered later, but it is best to work the programme out in advance.

Then, the individual blocks must be prepared. These blocks are composed of a number of segments, which could comprise:

* Moving footage.
* A still frame.
* A freeze frame.
* A step frame sequence.
* A slow motion sequence.
* A page of computer text.
* A page of computer graphics.
* A page of computer text over computer graphics.
* Computer text or graphics over a video frame.
* A page of teletext.
* Choice of audio tracks (sound, of course, may be included with many of these sequences).

This 'segment menu' will usually also include options which lead to branching, such as:

* Free choice, in which the user makes a decision from a list of available options.
* Test, in which the user is tested and then given appropriate feedback. (Various types of test are discussed in Chapter 36.)

If records are being kept in 'hard copy', the segment menu can offer to print out selected pages of text, or data the system has collected itself, such as student scores or a record of the ways in which the programme has been used.

Ideally, the authoring system is something between a prompt and a strategy, which helps individual authors who may not have experience in this kind of course preparation or information programming. It ultimately manages the author, the end user and information generated by both.

CHOOSING AN AUTHORING SYSTEM

The most important point in selecting any component of an interactive video project–whether the computer hardware or the presenter who introduces the programme–is choosing what is appropriate for the people involved. This means styling the content of the programme to its users, and the design structure to its authors. The best authoring system is the one with which the people who are going to use it feel comfortable. That is easy enough to say, harder to define. Ultimately, there is no clear answer–the decision must rest with the people who are running the project. However, there are some obvious considerations.

One is compatibility. At least so far, authoring systems tend to be designed only for certain specific computers. Currently, the Apple and the IBM PC (Personal Computer) have a wide choice of authoring software. If you are buying a computer, it is a good idea to include 'authoring software' on your checklist of important features. If you already have a computer, the company who sold it to you, or a good software distributor, should be able to advise you.

Three points on which to evaluate an authoring system are:

* Programme creation.

* Programme management.

* User management.

A trainer altering a program on the CAVIS system.

'Programme creation' describes the basic construction of a programme or lesson. 'Programme management' refers to the modification of existing programmes, through changes and additions, and the preparation of documentation which records how the lesson was constructed. (Documentation is essential if the lesson is to be understood by other authors or designers, or if it is to be modified in any way.)

'User management' describes the sometimes detailed records that can be kept to monitor either individual users or the use of the programme overall. The progress of any individual user can be recorded step by step, to show which segments were seen, which branching paths taken, how the user performed on tests, and even how much time was taken at each stage of the programme. With these records, tutors or trainers can analyse each user's performance, and offer tuition or alternative instruction as necessary.

Cumulatively, these results also monitor the effectiveness of the programme: if all users fall at the same fence, it is likely to be the programme that needs changing. The overall use of the system can also be monitored, as it often is in marketing applications, to show how often each programme was used, at what times of the day, and so forth. The system can be employed, too, collecting useful information from its users on their background, reactions and expectations. As a marketing tool, this is an especially painless way of conducting market research.

The authoring system should be 'user friendly'—that is, it should be easy and inviting to learn and to use. Its manual should be written in plain language, and the instructions on the screen should be easy to follow. Whether this is so should be clear from a light read through the manual or a typical menu on the screen. The idea is not to impress users with the complexity of the technology, but to ease them into programming through a series of simple operations using familiar language. The authoring system should work for its users, and not they for it.

An authoring system that comes with as much documentation as a nuclear power plant is not a practical option for people who want to make simple programmes quickly and cheaply. If the authoring system is very complex, it is no easier to use than a computer language. The point of an authoring system is to obviate as much implicit programming activity as possible, and free the author to concentrate on the preparation of the lesson or information.

How long individual authors take to come 'on stream' depends on several factors, but in many teaching or training situations it is reasonable to expect that authors should be able to make simple programmes on the first day, and adventurous ones with only a little experience.

The authoring system should offer designers and users scope. Basic branching strategy is now pretty much universal, but the modular construction of different systems does vary. Initially, and for some applications, simple programmes are fine. However, as authors and users grow accustomed to interactive technology, they should be able to work through more ambitious programmes. A creative designer can produce interesting programmes on limited means, but it will be frustrating to designers and users alike if the routine soon becomes restrictive.

Of course, there are limitations to the best of authoring systems, and new and better systems are emerging all the time. For example, at the time of writing, there was no inexpensive, efficient method way for authors to lay computer-generated text and graphics on top the video picture. Undoubtedly, authoring systems will continue to grow more ambitious and versatile for some years to come.

The short answer is, when the time comes for you to choose an authoring system, shop around, read the computer magazines, and ask the advice of a qualified consultant. Don't go for anything until you're sure of it, and do ask to see samples of work produced through the system. Authoring systems are definitely a great leap forward in bringing the benefits of computer technology to a wider market, but one which does not address the real needs of its users is worse than useless. The right authoring system and the right hardware are as important as the right design and the right visuals.

CHAPTER 23: COMPUTER TEXT AND GRAPHICS

Computer-generated text and graphics, like many of the topics discussed in this book, are a study unto themselves—and one which is only too quickly shoots off over the heads of most interested amateurs. Yet it is worth addressing briefly here. Since interactive video represents the convergence of video and computer technology, it is only reasonable to expect that the computer which controls the video programme should also generate complementary text and graphics. Ways of combining the two signals on one screen have been of vital interest to both computer and video manufacturers for some time.

The main obstacle has been the simple fact that computer and video screens are composed differently. The key letters in computing are 'RGB'. These describe the basic red-green-blue signal from which colour pictures are built up on both television and computer screens. Both employ a myriad of tiny dots, tightly packed in rows across the screen, each of which is lit individually when the picture is displayed.

The difference is that, in most television transmissions, the dots are grouped in long rows—the 'scan lines' described in Chapter 10—while in computer technology, each is a separate entity, called a 'pixel' (abbreviated from 'picture element'). The colour and intensity of each pixel can be specified precisely, and the ultimate effect, like that of tapestry or mosaic, is often deceptively like that of a liquid medium such as paint or ink.

Furthermore, the separate video and computer signals have to be synchronised. A video transmission is regulated by a series of rhythmic 'pulses' recorded along with sound and picture signals. These must be synchronised when signals from separate sources are mixed—to combine video and computer signals, for example, to superimpose captions, or to split a screen among two or more pictures. This is effected by means of a 'gen-lock' which locks the pulses generated on separate signals to a common beat. A gen-lock is needed to lay computer signals over video signals.

RESOLUTION

The quality of graphic you see displayed depends largely on the 'resolution' of the screen on which it is presented. This is dictated by the number of pixels on the screen. The higher the resolution, the more the pixels, and the finer the detail. Typical screens these days run from 300 by 200 pixels to 900 by 400, or 1500 by 1200. The upper limit—'very high resolution'—is currently around 2000 by 2000.

VECTORS, WIRE-FRAME, RASTERS AND BIT-MAPPING

Computer graphics can be approached from more than one direction, either through detailed programming or, like authoring systems, using software with shortcuts built-in.

There are two basic approaches to the design of computer- generated graphics. 'Vector graphics' or 'wire frame graphics' are made up of any number of fine lines, straight or curved. The shapes these outline can later be filled in with colour. With a great many fine lines, and subtle gradations of colour, vector graphics can be quite impressive. They often retain the 'feel' of computer-aided animation, but this can be to their advantage – high tech art for high tech applications.

'Raster graphics' or 'bit map graphics', on the other hand, are the high tech answer to petit point. You may remember the word raster from the description of the video screen in Chapter 10. In this system, each pixel is addressed separately, and the picture is built up point by point like a fine piece of needlework. The quality is often exceptionally high, but the alternative name, 'bit map', suggests the detailed programming – bit by bit – that can go into the preparation of one of these graphics.

Bit mapped graphics can also take up a great deal of storage space on the floppy diskette, as each one constitutes a screen full of information and instructions to the computer. A vector graphic, on the other hand, can be stored much more economically, with a single instruction describing each line or curve drawn. This is something to bear in mind if storage space on the floppy diskette is a consideration.

PRODUCING COMPUTER TEXT AND GRAPHICS

Computer graphics and text can be created in several ways.

* Some facility to design text for the computer screen is an essential part of any authoring system. Authors usually have a choice of fonts (that is, styles of lettering), size, letter colour and background colour, and can dictate spacing, centring and other factors which affect the appearance of the screen. Some authoring systems, like Apple's SuperPILOT, have their own graphics packages as well.

* Intricate or unusual graphics can be designed through conventional computer programming. This is far and away the most expensive way of producing artwork in terms of labour, time and money.

* Many video production or facilities houses have equipment specifically for text, captions and graphics. For example, Chyron IV ('not one of the lost kings of Egypt') is a character generator, while Dubner ('not so much a Czechoslovakian spy, more a full-function character generator with painting software and real-time animation')[1] is a real magic box which can turn plain artwork into 3D high tech wizardry.

Then there are graphics software packages. Like authoring systems, these vary enormously in scope and technique. A good designer can produce interesting work with virtually any tools, but it is not practical to devote hours of labour to a graphic which is beyond the scope of the software, or which could be produced more economically another way–on professional equipment in a production house, or on some other medium such as good old-fashioned ink on paper.

The decision to use computer graphics instead of traditional artwork is one of both style and economy. If a computer is already being used in authoring, graphics may be designed through a complementary software package. But don't forget that even with a simple software package, designing graphics can quickly become labour-intensive out of all proportion to the value of the finished product. Traditional artwork–drawings on paper–is often quite good enough, although computers can be a great help in executing animated pictures. It's very much a matter of horses for courses–and, too, of the advice of qualified commentators.

INPUT DEVICES

There are a number of ways of drawing graphics for the computer screen.

* Some graphics software packages employ the computer keyboard, giving special values to familiar keys to direct the 'cursor' around the screen and fulfil commands. (The cursor is the special character on the screen that tells the user–whether an accountant, a word processor operator or a computer graphics designer–where on the screen information can be entered next). In these systems, the cursor becomes a pencil point, or the tip of a paint brush, under the user's control. This kind of package is strong on geometric shapes, but poor on freehand drawing.

* Others use a 'light pen' which actually 'draws' on the screen.

* Some packages move away from the screen onto a 'graphics table' or 'tablet'. This is a sensitive drawing board on which users can block or sketch designs using some form of pen, or a 'mouse' or 'turtle'. This zoomorphic imagery describes small, rolling remote control devices which can be moved smoothly over the graphics table, their movement describing the movement of the cursor on the screen.

Again, the best choice is a combination of hardware which is compatible with the rest of the system, and economical in proportion to the rest of the project, together with software which is equal to the requirements of the job, and with which the designers feel comfortable.

TELETEXT

The development of teletext is germane to the integration of video and computer technology. Etymologically, 'teletext' is a miscegenation of Greek and Latin, presumably meaning 'printed words from afar'. Teletext provides the pages of news and public information which are now a common feature of broadcast television. The teletext signal occupies part of the previously-underemployed 'blanking lines' which separate each frame of video from the next. An ordinary television 'with teletext' can be used to intercept this signal for non-broadcast applications, to generate captions (in different sizes and colours, but, so far, only one font) as well as simple block graphics.

TEXT AND GRAPHICS OVERLAY

As we've said, once the computer is a captive element of the interactive delivery system, it may as well be exploited to the full. The idea of combining the two signals, so that computer-generated text and graphics can actually be laid over relevant still or moving footage, is an obvious one. However, 'graphics overlay', as it is called, is still easier said than done. Companies from the largest to the smallest have worked on it for years, and the fruit of their labours is only now coming onto the market.

There are currently three main ways of laying text and graphics over video:

* Video 'keying' is standard technology in video and television studios. It is used to superimpose captions ('supers') over video footage, and to produce special effects such as slipping the face of the weather forecaster into a corner of the weather map like that of a cherub in a painted ceiling.

* Then there is teletext overlay. It took some time to perfect, and is still rather limited aesthetically, but is a cheap and easy alternative in delivery systems which include a teletext encoder of the kind now standard in many television sets.

* Ultimately, there are systems which simply convert video signals to RGB. 'Video reprocessing' was an early high tech answer to computer/video interface, and a couple of packages were developed in the US to put NTSC video signals on a computer screen. As a feature of the newest authoring systems, 'overlay' is another step toward the integrated technology of the future, in which the components of many media are brought together in one package.

None of this, of course, addresses the question of screen design (what to do once you have the tools). That is a conceptual rather than a technical question, and is discussed in Section VI. Between here and there lies some more hardware: the delivery system.

1 Roger Holdsworth and Rob Oliver, 'Briefing on Video', Direction, September 1983.

CHAPTER 24: DELIVERY SYSTEMS

The term 'delivery system' describes the equipment on which an interactive video programme is actually presented. A typical configuration draws on both video and computer technology, and sometimes on elements (such as interface packages) designed specifically to effect interactivity.

This should not be the first chapter you read. Nor should you read it with a cheque book by your side. We appreciate the temptation to get your hands on the hardware first and read the instructions later. But we still recommend that before you spend any money, you study the whole book carefully, and then go out and discover what is actually on the market at the time you want to buy.

So, with these qualifications, let us consider what a basic interactive video delivery system comprises:

* A video player, tape or disc.

* At least one screen, video and/or computer.

* Some sort of remote control device (typically, the video player's keypad, or a computer keyboard).

* Computing power of some kind, either a microprocessor built into the video player, or an external computer (mainframe, mini, or – and usually - micro).

The choice of hardware affects the design of all the programmes presented through the system. Initial decisions depend on the fundamental choice of medium (whether tape or disc) and on the level on which the system is to operate (what sort of computing power is to be employed). If you have decided to base the system on equipment which you already own, the decision is how best to build around it. If the project is a one-off, the delivery system can be tailored to the needs of one programme. More likely, the hardware will be acquired with a view to making many programmes, well into the future, and decisions taken now must weighed with that consideration in mind.

However, the designers and manufacturers of both video and computer equipment have realised that compatibility is in everyone's best interests. More and more, communication between disparate pieces of equipment is becoming standardised. So, if the system is based on standard equipment and packages, it is generally fairly easy to add a screen here, or upgrade a player there.

There is no standard combination of equipment with which to present an interactive video programme. The smallest configuration currently available comprises:

* An industrial standard videodisc player with its own onboard microprocessor.

* The videodisc player's own remote control keypad, or some purpose-built equivalent to this.

* A colour video monitor.

In a system of this kind, the whole interactive programme – sound, pictures, and instructions to the microprocessor – is encoded on the videodisc itself. When the disc is put into the player, the program instructions are loaded into the microprocessor's own small temporary memory. The player can then run the video programme interactively.

With domestic standard equipment, this would probably be a CAV or 'active play' disc such as 'Teach Yourself Tennis' or 'Car Maintenance' or a 'MysteryDisc' game. Not all domestic standard disc players are equal even to this low level of interactivity, nor are all the discs compatible with all types of machines. Consult the manufacturer's literature, and look for features like 'random access' and 'frame display' in the videodisc player. Check the fine print on the back of the sleeve to be sure of the disc.

With industrial standard equipment, this configuration is used in many marketing exercises, and in teaching and training courses where a stable body of information can be presented in a relatively straightforward way.

In terms of the scale introduced in Chapter 5, domestic standard videodisc players represent Level 1, and the industrial standard ones, Level 2. Both these use a minimal amount of hardware, are relatively cheap to buy, and easy to set up and transport. The simplicity of using just a disc and a simple remote control device like a keypad is appealing in many applications. Level 2 offers sufficient scope for projects in many fields. Industrial standard players now carry microprocessors of up to 5K RAM, and the disc can hold any number of programs. In fact, these load themselves onto the microprocessor, one after another, as the disc plays and without distracting the user in any way.

However, the whole programme – video, audio and instructions to the computer – is encoded on the disc itself and, once recorded, cannot be changed in any way. And, of course, a system of this kind is so far restricted to a limited number of videodisc players. The transition to disc is not one all video users are eager to make, especially if there is already a heavy investment in videotape hardware.

A system operating at Level 3 stores video programmes and computer programs separately, and can update or amend data stored in the computer – the control program itself or computer-generated text and/or graphics – without touching the video programme. Most videodisc players can be used at Level 3 and, at least so far, all tape systems (lacking onboard processing power) are obliged to work there.

A typical Level 3 configuration comprises:

* A videotape or videodisc player.

* A computer.

* At least one screen.

* Some form of input and/or control device.

* An interface package, if necessary.

The videotape machine can be a player or, preferably, a recorder/player, of either domestic or industrial standard, using either ½″ or ¾″ tape. Industrial standard ½″ VHS or ¾″ U-matic equipment is common in professional use.

At the time of writing, available videodisc systems are effectively playback only–the recordable disc is still an expensive tool for static information storage. So far, most work has been done with the laser disc, but now that the rival VHD and CED systems have interactive players on the market, attention is being focused on these systems, too.

Although many different computers can be linked to video equipment, most applications employ a micro (or personal, or home) computer. Basic peripheral devices usually include:

* One or two disk drives.

* A screen (preferably with an RGB colour monitor).

* A keyboard on which authors can prepare programmes and users interact with them.

* A printer for documenting the design of the system, and recording individual users' performances.

This may be enhanced to encompass:

* Control devices such as paddles and joysticks.

* A touch-sensitive screen or light pen.

* Aids to graphics-generation, such as a graphics tablet.

Two screens might be employed–one for video signals, and one for computer text and graphics. Alternatively, with the graphics overlay facility described in Chapter 23, the two signals could be on an RGB screen or a video monitor.

Control devices also include the video player's remote control keypad, purpose-built devices, and the computer's keyboard, paddle, joysticks, bar code wand, turtle or mouse. (Confused? Check the Glossary!)

Whether an interface package is needed is a matter of what equipment is used. The whole subject of interface is addressed in Chapter 21–but, again, the only way to be sure is to ask advice of qualified consultants as and when you are considering setting up the delivery system.

As we've said, if a computer is employed to control the video player, it might as well be exploited to generate text and graphics as well. This is particularly useful if any of the material in the presentation is 'volatile'—that is, likely to change regularly or dramatically over the course of time. Stored in the computer, it can be updated or amended at any time, without affecting the video recording. Furthermore, if what is recorded on the video changes at some later date, text or drawings explaining these changes can be generated through the computer. If, for instance, a piece of equipment were modified a year after the video material was shot, a line drawing illustrating the modification, and a short piece of explanatory text, could be generated through the computer to explain this. With an overlay facility, computer-generated text or graphics can even be laid on top of the video picture.

So, the delivery system could comprise as little as a videodisc player and its accessories, or—in a simulator, for example—could stretch to several video players, a mainframe computer, any number of screens, and peripherals of many descriptions.

With the emergence of computer/video interface packages and authoring systems of the kind described in Chapters 21 and 22, the task of setting up an interactive video delivery system from disparate computer and video hardware can be relatively cheap and easy. With these packages, computer and video equipment acquired at different times, to fulfil separate functions, can be linked to make an interactive video delivery system. The convenience, and economy, of using hardware already on hand obviously has a strong appeal to both business and home users.

THE INTERACTIVE VIDEO CENTRE

There are many differences between interactive video programmes which have to appeal to a general public, as point-of-sale units do, and those which enjoy a captive audience, as do most training programmes. Questions of approach, style and content aside, one striking difference is in the actual presentation of the hardware.

Point-of-sale units are invariably housed in display cabinets which have the multiple function of protecting the equipment, attracting passers-by, and instructing consumers in the use of the system. Simulators by definition are designed to look as much as possible like the real thing, be it an aeroplane cockpit or the driver's seat of a compact car. Only too many teaching and training packages are presented with hardly a thought to the display of the hardware beyond ensuring that the student has visual or manual access to the relevant hardware.

The design of a good display unit is an extension of good programme design. There's little use in having a wonderful marketing programme if you cannot persuade customers that it is fun and easy to use. It is plain foolhardy to put delicate, expensive equipment into general use without some protection from inquiring fingers. Virtually any application benefits from some creative and practical approach to the display of the hardware. This is especially important if the technology is likely to be new to the people being asked to use it. You need only wander around a computer show with an eye to the display stands—many of which are quite elaborate—to see how important good presentation is to the impression a machine may make on a first-time user.

The ideal display unit is sturdy, compact and attractive. It puts the bits to which people need access – screens, control devices, disc drives and so on - within easy reach. It probably conceals the parts the user doesn't need, either to modify the impression of too much high tech or to protect the equipment from dust, interference and vandalism.

Such a unit is often the work of a someone commissioned especially for the job, quite possibly a professional display designer. Even in the most modest application, a little attention to the display of the hardware will go a long way to helping users make the best use of the system.

If you have a whole room dedicated to the system, it could be set up as an Interactive Video Centre, with an attractive delivery system and separate 'quiet' study and rest places. If there are pieces of equipment which the user does not need to see or touch – if, for example, the tape or disc is loaded by a supervisor – that equipment and its attendant yards of cable could be concealed. Even a piece of baize over the table (hiding the machines and wires underneath) and a comfortable chair is an invitation to the user. In presentation as in other aspects of the design of the programme and its delivery system, attention to detail can make all the difference to the way in which the technology is received and employed.

CHAPTER 25: CHOOSING THE RIGHT FORMAT

Let us first of all consider what video medium the delivery system is going to be employ, whether tape or disc, and which format—whether ½″ or ¾″ tape, laser or capacitance disc. Remember, there's more to interactive video than broadcast-quality pictures, and not all video systems adapt equally well to interactivity.

To work well in interactive technology, a video system must offer at least the basic features:

* Good quality pictures and sound.
* Accurate and rapid random access.
* Easy and quiet operation.

Let us consider these points individually.

PICTURE QUALITY

Picture quality is influenced not only by the choice of medium, but also by the quality of the screen on which the picture is displayed and the quality of the source material. The question of screens is considered later in this section and that of production techniques, in Section VII.

However, some systems offer inherently better picture quality and stability than others. For one thing, wear significantly reduces picture quality. Laser discs, sealed within a tough protective shell and having no direct contact with the reading device, are the most stable medium; tape—and, to a much lesser extent, capacitance discs—both wear a little every time they are played. Laser discs have their topcoat, and capacitance discs and video cassettes both come in protective cases. However, all video media are vulnerable to damage and wear, and all should be treated with care and respect.

These considerations aside, picture quality is good on all three disc systems. Among the tape systems, U-matic, the ¾″ 'institutional' standard, offers the best picture quality (PAL high-band tapes are broadcast standard). It is a good tape to edit—to the third generation at least—and its players are sturdy and reliable. But while newer players are smoother and quieter than earlier models, both cassette and player are bulky, and the system is slow-winding, with sharp mechanical noises at every stop and start.

There is an argument for shooting and editing on either ¾" or professional standard 1" tape, and then replicating the finished programme onto a smaller, tidier ½" format in the final stages of production. The tapes do wear, but as they are fairly cheap and easy to produce, they can be replaced as necessary. Industrial standard players are made for professional use, and are sturdier and often more versatile than their domestic equivalents.

SOUND QUALITY

Sound has never been a strong point in broadcast television, nor in non-broadcast video. You usually need elements of a separate sound system to get good stereo reproduction with a television or video player–which is not always a practical suggestion for anything beyond a home entertainment centre.

Of course, for most interactive video applications, it is not stereo *per se* but two separate audio tracks which are important. If the two audio tracks can be recorded and controlled independently, the programme could offer a choice of soundtracks–in two languages, say, or at two levels of comprehension. As with many other features, the two audio tracks are standard in new, industrial quality machines, but are not always found in older players, nor in all domestic standard equipment. However the audio facility is described, the important feature is that the two channels are absolutely independent of one another.

In some tape systems, the second audio channel is needed to record time codes and other control information vital to the interactive programme. This reduces your options, but is necessary if the system is to work interactively at all.

In the present generation of video players, one drawback to the use of special effects like slow motion, step framing and freeze frame, as well as to still frame storage, has been the loss of the audio track linked to footage moving at normal speed. Developments in video sound technology which seek to address this problem–offer features like compressed sound, digitised audio and video signals, and 'sound over stills' converters.

(A converter is a device that stores audio signals as though they were video signals–very compactly. The audio still frame converter marketed by EECO (Electrical Engineering Corporation of California), for example, can store up to 40 seconds' audio with a single still frame on disc or tape, with a total of up to 120 hours' audio and 10,000 still frames on a single laser disc. Sony offer a similar package with their NTSC videodisc player. The VHD (Video High Density) disc system has an optional AHD (Audio High Density) adaptor which fulfills a similar function.)

One option immediately, and cheaply, available to people who want to tie unlimited audio to stills and moving footage, is an interface package which links a conventional audio tape recorder to a videodisc player. The audio tape records both the soundtrack and commands to the player. Refer back to Chapter 21 for more information on interface packages.

FRAME-ACCURATE RANDOM ACCESS

Quick and accurate random access is a function of both the precision with which individual frames can be 'addressed'–that is, found exactly in a 'search' through the video material–and the speed and efficiency with which the player actually stops and starts, searches and finds.

In Chapter 14 we saw that certain laser discs are 'frame addressable'–that is, any single frame can be requested, sought, found and brought to the screen within seconds, whether a still or the first frame in a sequence of moving footage. These are the CAV (constant angular velocity) or 'active play' discs.

It is the ratio of one frame per revolution that makes CAV discs so accurate. Of the capacitance systems, VHD has two frames per revolution, and CED has four. This makes it more difficult to guarantee frame-accurate random access, and, as we will see shortly, to offer other interactive features. Some tape systems are addressed in terms of chronological time–minutes and seconds–rather individual frames. When there are 25 or 30 frames in one second of video, this can mean the difference between the last frames of one moving sequence and the first frames of another, or one still frame and another, utterly unrelated one.

More accurate tape systems employ 'pulses' recorded on the tape to identify each frame. The videotape controller in the interface package picks up these signals, and the computer uses them to direct searches. However, the tape player itself often winds slightly faster or slower at one time than another–so, after a few searches, the tape is starting too late or too soon. For this reason, the synchronisation of tape-based systems usually has to be watched closely, and set up afresh many times in a day to keep the tape player running in tandem with its controller.

EASY, QUIET OPERATION

The quick and quiet running of the player itself contributes both to efficient random access and to the popularity of the system with the people who use it. The object is to avoid long pauses and distracting mechanical noises while the machine stops and starts, searches and shifts.

In practice, users often appreciate a break between scenes, to collect themselves and to prepare for the next segment. In a Level 3 application, the computer can generate text or graphics to fill the screen while a search is going on. However, the ideal machine still moves so quickly and quietly that the user's attention is not lost or distracted while waiting for the next segment.

The disc systems move smoothly and quietly between non-consecutive scenes. Access is virtually instantaneous in laser disc systems, in which there are no mechanical parts to disengage when the programme moves from one segment to another, and in the VHD system, in which a sensor glides lightly over the surface of the disc. But even the first-generation interactive CED players, in which a stylus actually rides a groove in the disc's surface, 'worst' access time is still only a matter of thirty seconds.

Tape is slower than disc, mainly because the tape has to wind from one reel to the other. (Sometimes, too, the tape player fails to find the right frame on the first pass, and has to re-wind and try again). The speed and accuracy with which any one frame can be identified facilitates a faster search, but so does the speed with which the machine can cue or re-wind the tape. A system which plays linear tapes well enough will soon become maddeningly distracting in an interactive application if it clicks and clunks and stops for little rests all along the way.

The phrase 'worst access time' describes the longest possible journey from one place to another on a tape or disc. In practice, with good 'real estate' management, material is arranged to minimise the length of this journey. (For example, RCA recommend placing the main menu, or table of contents, in the middle of the CED disc: this reduces maximum journey time from the menu to any segment to fifteen seconds.)

All these features – sound and picture quality, quick and accurate random access, easy and quiet operation – contribute to that ultimate goal, 'user friendliness'.

USER FRIENDLINESS

The key to the success of any project, whether the audience is a captive one of trainees or an elusive one of passersby in a busy shopping precinct, is engaging and retaining the attention and co-operation of the user.

The term 'user-friendliness' comes from computing. It handily sums up in two words the vexed question of how the myriad individual components of a delivery system of any kind ultimately appeal to the people who have to work with it. That appeal depends on both hardware and software, and is as much a question of style as engineering.

The simple physical appearance of the system is of critical importance in creating a favourable impression. The general trend in the design of videodisc players is toward spare, clean lines and broad, flat surfaces. The look is very much that of high technology – which some people find attractive, and others, intimidating. Some videotape players are chunky, some are compact: ½" tape machines are smaller and generally more attractive than the ¾" models.

In Chapter 26, we will begin to look at remote control devices such as keypads, light pens, touch-sensitive screens and so on. Of these, the keypad is the only one that usually comes with the video player: if the keypad is to be part of the delivery system, its design, too, is important.

Of course, in many applications users see no more of the hardware than the screen and the device with which they control the programme – a keypad or light pen, for example. The rest is housed within a purpose-built display unit which both protects the equipment and serves to attract and instruct people using the system. The idea of designing a display stand was discussed in the last chapter.

LOOKING FOR THE BASICS

The delivery system must draw on equipment which has the potential to work interactively. The critical consideration here is the video player, which must be either

Level 2 and ready to stand alone, or at least Level 1 and capable of linking to an external computer.

We introduced Levels of the Nebraska scale in Chapter 5. Although devised to describe interactivity in videodisc players, the basic features are those expected of any interactive video delivery system. Once again, they are:

* random access (search)
* two audio tracks
* remote control
* forward and reverse
* scan
* slow motion
* freeze frame
* step frame

We have touched on the question of rapid and accurate random access, and that of audio facilities. Remote control devices are a feature of the delivery system as a whole, and may be standard video or computer peripherals (discussed in Chapter 26), or purpose-built tools. Level 2 often employs the industrial standard videodisc player's own keypad, while Level 3, as we've seen, can use just about anything. Forward and reverse scan and slow motion are common features of most modern players, tape and disc, although you could make an interactive programme without them.

Step frame and freeze frame both involve holding a single video frame on the screen for an indefinite period of time. The use of still frames depends on this. CAV laser disc can hold any frame for any length of time without damage: this is a function both of the ratio of one frame per revolution and the fact that there is no direct contact between the reading device and the surface of the disc.

As we saw in Chapter 15, at time of writing, neither of the two capacitance systems, VHD and CED, can freeze satisfactorily on any frame at random, and neither likes to hold a still frame for any length of time. Later generations of VHD players promise to overcome this, but until then, this is a drawback in capacitance systems.

Similarly, no videotape system so far can hold a still frame for more than a few minutes before the player itself must reject the tape to avoid damaging it. The tape is held still but the video heads continue to rotate, reading the same, stationary video tracks over and over. This is not how videotape players are meant to work, and the still frame often suffers vertical 'judder' and a horizontal band of interference. Furthermore, holding the tape in this way contributes significantly to the deterioration of the sensitive emulsion on which the video signal is recorded.

PRODUCTION CONSIDERATIONS

On technical points of the kind discussed here, laser discs score the highest, with VHD running a close second. However, there are other factors in the selection of the video equipment for an interactive delivery system.

Many organisations prefer to produce their video material entirely in-house – for security reasons, perhaps, or to capitalise on established production facilities. This control must be relinquished when videodiscs are involved, for, so far, all videodiscs must be manufactured under strict conditions on expensive equipment, in one of a still relatively few disc pressing plants. Then, too, while the actual replication cost of discs can be small when they are pressed in large numbers, the unit price per disc can be extremely high when only small quantities are required.

Of course, many people already own videotape equipment and production facilities, and have expertise in traditional tape programming. In a large organisation, interactive video may be only part of a larger tape-based communication system. Even if it is an important part, this does not necessarily justify the introduction of a discursive element into a successfully integrated network.

Some people, for a variety of reasons, simply prefer tape as a medium. There are also arguments for interactive sound/slide systems – and for computer-based technology which employs only its own text and graphics. There is room for all, and good examples of interactive technology at all levels. The ultimate decision must be made considering the delivery system as a whole.

CHAPTER 26: DELIVERY SYSTEM HARDWARE

There is no standard combination of equipment with which to present an interactive video programme. The basic requirements are outlined in Chapter 24: a video player, computing power of some kind, a screen, a control device. This can vary enormously–the three-piece configuration of a Level 2 system (videodisc player, keypad, monitor) may be the minimum, but in Level 3, the sky's the limit.

What we plan to do here is simply to explain some of the components of an interactive video delivery system. Someone who already has a video system and/or a computer may find this information painfully obvious–it is intended for people new to one field or the other, or both.

INPUT/OUTPUT DEVICES

This heading covers a wide range of devices which the computer would regard as peripheral, in that they are all under the command of its central processor, but which people tend to regard as central, in that they are the tools which communicate between people and the computer. This seems a reasonable place to start our quick look at hardware.

Input describes the transfer of information from some medium–manuscript, audio tape, the user's head or any other thing or place–into a computer. Once in the computer, this information, no matter what the form in which it was input, is converted into the electrical codes which are the computer's native language. The computer processes and stores all the information it receives in this way. Only when the information is retrieved for use by the computer operator is it converted back into a form which people can understand and use. Output describes this retrieval process, and the display of data converted from computer codes into some familiar mode of human communication, such as words or pictures.

In short then, data is entered into the computer through an input device, converted into electrical codes by the computer's own software, processed within its central processor, converted back into a form which people can easily interpret, and displayed on an output device.

Typically, information is input through the computer's keyboard and projected onto its screen, processed within its central processor, and stored on a magnetic floppy disk. It can later be retrieved and output through a display on the computer screen and/or a printout onto paper. The one thing which may be clear in this apparently tangled web is that some of the basic hardware comes into play at several different stages in the programme.

In an interactive video application, the programme-makers need more equipment to design the programme than its users do to run it. The programme-makers often have to design and record both the instructions used by the computer to run that programme, and the computer-generated text and graphics which appear in it. This is usually effected through a keyboard containing both character keys bearing alphanumeric and other symbols, and keys dedicated to specific computing functions. If computer-generated graphics are to be included, they may be designed through this keyboard or with some other tool.

The user, too, may need a keyboard, although probably one with fewer keys. On the other hand, the user may need only a simple keypad, a touch-sensitive screen, a light pen or other such device to run the programme and to answer questions: increasingly, other, more versatile, means of communicating with the computer are being explored. Other tools have their uses, but keyboards are still integral to delivery systems of the kind used in interactive video.

THE COMPUTER KEYBOARD

The computer keyboard is similar to that of a conventional typewriter, but has, in addition to the keys representing alphanumeric and other familiar symbols and characters, rather more keys dedicated to specific functions. Even the humble manual typewriter has dedicated (or 'global') keys such as Tab, Back Space and Shift. An electric typewriter has more, an electronic typewriter still more, and a word processor often as many as a full-fledged computer (a word processor is, after all, a type of dedicated computer).

A keyboard can be used by a programmer to give instructions to the computer, to record text to appear on the screen, and to design computer-generated graphics. It allows the user to interact with the computer, to make selections and to respond to questions.

The design of the keyboard, with respect to both the number, function and arrangement of the keys, is part of the design of the computer. In some systems, it is possible to design a keyboard to address special needs of the programme. It is always worth paying attention to the design of the keyboard, whether on a typewriter, word processor or computer.

The keys should be clearly marked, so you can tell at a glance what any one does. They should be arranged in discrete blocks, with the various groups of keys clearly separated and neatly arranged around the main block of alphanumeric and other character keys. This makes it easier to find groups of related keys (those which execute similar functions), and to avoid hitting a wrong key by mistake.

Although basic keyboard skill–'typing'–is increasingly being recognised as a valuable asset in jobs other than secretarial ones, many computer users, especially casual ones, are still two-finger typists. A well-designed keyboard makes the user's job much easier, and reduces the incidence of errors. Good courseware may allow for typing errors on the user's part, but for the programmer, errors can have far more serious implications in computing than in copy-typing. The most important thing is that the person who most often uses the keyboard should be comfortable with the look and feel and functioning of it.

OTHER INPUT DEVICES

A keypad is simpler than a keyboard: it typically comprises a block of alphanumeric keys, plus keys dedicated to various relevant functions. In some interactive delivery systems the keypad is the video player's remote control device, with ten numeric keys, and keys to control rapid random access, and the fast and slow forward and backward motion of the video. The keypad is often designed to sustain a lot of indiscriminate use by casual users, and so contains an minimal number of keys, simply labelled and cased in a tamper-proof shell.

A touch-sensitive screen responds to the touch of a finger. Typically, the user may execute certain basic functions (such as starting, stopping or suspending the programme) at any time simply by stroking the appropriate part of the screen: a diagonal stroke across the upper lefthand corner may, for instance, return the user from any segment in the programme back to the main menu. At the same time, responses to questions and choices can be made simply by touching the spot on the screen where a word or picture represents the desired option. In a maintenance manual, for example, a photograph or drawing of a piece of equipment may introduce a choice of segments describing its various component parts; the user simply touches a specific part on the picture to receive detailed information about it.

A light pen looks like the familiar penlight style of flashlight, and works in conjunction with a cathode ray tube – the picture tube of a video or computer screen. It can be used by the programmer both to design and to amend information (usually, graphics) on the screen, and by the user to make choices and answer questions. The user may be presented with a basic drawing, for instance, and be asked to sketch in some supplementary information.

Paddles and joysticks are devices familiar from video and computer games. A paddle has a knob, a joystick looks like the stickshift of a car. They can be used to guide a cursor (the distinctive shape that marks the user's place on the screen); alternatively, they can be used as dummy controls in simulation exercises such as those conducted to train equipment operators.

Of course, as simulation exercises demonstrate, peripheral control devices can include just about anything that can be wired to communicate with the system, from the American Heart Association's 'manikins' (described in Chapter 4) to the instrument panel of an aeroplane in which every switch, dial and gauge works realistically. Bar code wands, like those used to tally purchases in computerised shops, can be used with specially coded notebooks to do exercises in a training course, for example, or to order goods from a point-of-sale unit.

OUTPUT DEVICES

RECEIVERS AND MONITORS

Interactivity is currently restricted to pre-recorded tapes and discs, and probably will be for some time yet: viewer-participation broadcasting is still a highly speculative business. So it is not the 'off-air' TV receiver, but video monitor and computer screen which concern us.

The TV receiver uses an antenna to intercept UHF and VHF (ultra and very high frequency) signals modulated for broadcast transmission. The monitor receives unmodulated, high quality signals transmitted at low frequency directly from the video tape or disc player. Many domestic video players and microcomputers can receive a broadcast signal, and most computers can also tie in to a telephone line through a device called a modem. This allows the computer to communicate with others, and to use public information – like teletext or databases – easily and cheaply.

For versatility, a combination receiver/monitor is a practical choice of equipment for many people, even in industrial applications. These days, a receiver with a teletext encoder is both the key to a whole field of information, and a tool for character generation.

SCREENS

Traditionally the screen has been a vital but passive component of the delivery system, necessary to transmit the visual display, but of no use otherwise. The introduction of touch-sensitive screens and light pens has given the screen an active role within the delivery system that presages the development of ever more inventive peripherals in the near future.

In assembling an interactive video delivery system from scratch, the choice amongst screens is one of several basic decisions. However, it is possible to upgrade or expand a delivery system piece by piece, even one assembled as a package by outside specialists. As we've said, hardware designers and manufacturers realise that compatibility is in everyone's best interests, and it is fairly easy to upgrade an system built around well-known makes of equipment.

It many applications, it makes sense to have two separate screens: one for video, one for computer-generated text and graphics. Information storage and 'encyclopaedic' programmes – ones in which a quantity of computer-generated textual commentary accompanies every video sequence – can be as awkward to present on a single screen as they are attractive on two separate screens. The use of two screens often has the comfortable feel of a book, with text on one side and pictures on the other. It is difficult in publishing to synchronise illustrations (especially colour plates) with their complementary text. But with two-screens in a delivery system, it is a routine matter to have every block of text illustrated with exactly the right visual material.

The Philips VP835 'Professional LaserVision' PAL-standard videodisc player, with plug-in ROM cartridge, keyboard and touch sensitive screen with teletext overlay.

As we explained in Chapter 23, the computer screen is a different beast to the video screen. There we discussed the development of 'overlay' systems that mix video and computer-generated signals on one screen. Developments such as this point to a shift in computer-based systems from the conventional video screen to the RGB screen.

PRINTOUT DEVICES

A variety of equipment has been developed to transfer information from the computer to paper. A printer is important in an interactive video delivery system both to document the design of the programme, and to record information such as student scores in a training programme, or to print receipts on a point-of-sale unit in a department store.

There is usually a choice between speed and quality in choosing a printer, but interactive video applications are generally less concerned with speed than quality. There are various kinds of printer, but the two which concern us are:

* Daisywheel printers, which are the most common type of formed character printers (those which use a changeable printhead in which each character is individually formed). They produce typescript in a variety of styles and sizes, at speeds averaging from 40 to 60 characters a second.

* Dot matrix printers, which form characters as patterns of tiny dots, rather than in typescript: the variety and delicacy of the characters they produce depends on the design of both the printhead and the computer program which controls it. They typically work at speeds from 30 to 300 characters per second, producing sharper print at lower speeds.

Plotters produce graphic information, usually as a graph plotted between vertical and horizontal co-ordinates. Dot matrix printers can also print out some computer-generated graphics.

There are some very impressive printers indeed now on the market, but a small, simple, quiet printer for occasional use is all that is usually needed in an interactive video delivery system.

COMPUTER STORAGE MEDIA

In current practice, the storage of computer information usually employs magnetic tape or disk. The punched card, which a decade ago really did seem like taking over the universe, has been largely supplanted by these more economic and versatile media, which communicate more directly between the programmer and the computer.

Magnetic computer tape is similar to ½" videotape, but stored on open reels rather than cassettes. Magnetic computer disk comprises a multi-layered plastic disk, with magnetic coating on its surfaces. Both are recorded and played back in essentially the same way as magnetic video and audio tape described in Chapter 11.

Everyday work on a microcomputer employs the 'floppy' disk – a thin, flexible disk sealed inside a firm plastic envelope. It is played within a special unit called the disk drive.

Microcomputers generally have one or two disk drives, which are usually independent peripherals. Mounted in the disk drive, the disk rotates within its envelope; windows within the envelope allow the reading heads access to the surface of the rapidly spinning disk.

The floppy disk has limited storage space. For this reason, some computers incorporate a rigid, or 'hard' disk (such as the Winchester disk), sealed within the body of the machine itself, which can be loaded with information from several floppy disks to serve as a large temporary memory to run long or complex programs. This is not the best contemporary technology can offer, for, as we suggested in Chapter 17, the unwieldly combination of video tape or disc, floppy disk and hard disk could be swept aside by the versatile, capacious recordable optical disc now entering the market.

In the meantime, the floppy disk is compact and inexpensive – but not indestructible: it can be damaged by careless handling and does wear out. A floppy disk must never be bent, nor subjected to any pressure (never use anything but a moist felt-tip pen to label a floppy disk); extremes of temperature, and pollution from fingermarks, tobacco smoke or dust can damage the disk, while exposure to a magnetic field can alter its contents irrevocably. It should, in short, be treated with the same respect as a videodisc.

Magnetic tape is usually seen as a long-term storage medium, and floppy disk as practical for medium-term storage, while the hard disk or the chip within the computer are used as temporary memory to store important information while the computer is working. (The chip is still commonly called the computer's 'core storage', even though that term properly refers to an older type of magnetic storage device.)

In an interactive video programme, computer information is usually kept on a floppy disk which ties into video and audio information recorded separately on magnetic tape or optical disc. As we said in Chapter 21, a separate audio tape may be used to record both the soundtrack and instructions to the computer. Of these, all but the optical videodisc are based on magnetic recording technology, and can be fairly easily amended or updated. It takes less time to retrieve information from computer disk than from computer tape, just as it takes less time to retrieve data from videodisc than videotape. Of course, recordable optical disc technology may sweep all these considerations aside just as magnetic storage media themselves once made short work of paper tape and punched cards.

We have not mentioned manufacturers, makes or models. Nor have we discussed 'buying a computer' as a topic. As we said in the beginning, this is not a trade magazine, but a general introduction to the ideas behind interactive video technology. When it comes time to buy, the only thing to do is to look at the hardware market both for what is available and what is just on the horizon, to analyse your own needs honestly and in detail, and to ask for qualified advice.

CHAPTER 27: DELIVERY SYSTEMS OFF THE SHELF

We have talked about videotape and videodisc players, keypads, screens and monitors, computers, keyboards, interface packages and authoring systems. Perhaps it is beginning to sound as though there is no easy route to interactive video except through this obstacle course of disparate hardware, software, cables and plugs. This is not so. To suggest that, before you can make programmes, you have to build a delivery system, is rather like starting a recipe for rabbit stew with ''First, catch your rabbit.''

There are in fact a number of interactive video delivery systems which you can buy right off the shelf. In Chapters 21 and 22, we described interface packages and authoring systems that make it easier to bring together standard video and computer equipment. Those tools are of use only if you already own, or are about to buy, the appropriate hardware. The packaged systems we are discussing here are one step on from that: they give you everything you need, from the authoring software down to the plugs, to set up an interactive video delivery system from scratch, to answer the needs of both the programme-makers and the end user.

Cobbling together a delivery system from disparate pieces of hardware, and selecting the software to complement it, can be economical and practical – even, if you are that way inclined, good fun. How practical and economical depends on whether the system is built around equipment already at hand, and on how well the people buying the new components know the market. However, if you have none of the basic hardware, and would rather read Das Kapital from cover to cover than touch the current issue of Which Micro? magazine, there is a certain appeal in the idea of a buying a complete system in a box.

For one thing, such a package means that you can just about buy the delivery system off the shelf today, and start to make programmes tomorrow. One decision and one supplier cuts through a great many of the problems of setting up, and obviates the usual compatibility problems.

Furthermore, the various pieces of equipment will look like one comprehensive unit, rather than a motley collection of disparate hardware. The discrete components of the delivery system will all reflect the same lines, colours and motifs in their external design, and may even come housed in a stylish display unit. The effect of this on everyone who works with the equipment, either as a programme designer or an end user, is important in establishing good relations between the system and the people who use it.

Most off-the-shelf packages have been designed for teaching and training or for marketing applications. While some attempt to be versatile, there is a fairly sharp distinction between the two.

Two systems aimed at the marketing end of the industry illustrate the range of packages available.

* VCM (Video Communications/Merchandising) Systems, an American company, were mentioned in Chapter 7. Their product is an upmarket point-of-sale unit to which retailers 'subscribe' rather than make their own programmes. The system uses videodisc and comes housed in a substantial purpose-built display unit that looks rather like a juke box. As with a juke box, customers use a combination of lettered and numbered keys to request selections. The unit is designed around a specific application such as, for example, the Greenday Video DIY and gardening 'magazine' cited in Chapter 7.

* Video-dex, 'The Master Controller', by Videodetics is, on the other hand, a versatile tape-based system which aims to cater for virtually all markets. The emblem of the system is its keypad, a freestanding unit with four lettered and twelve numbered keys. The rest of the hardware can be presented in any way the customer chooses. The system is compatible with Betamax, VHS and some U-matic tape players in NTSC and PAL. It has found a variety of uses in marketing and training.

Clearly, these systems vary considerably in design, cost and versatility. This is as true in training as in marketing, for a delivery system can cost anything from £2,000 to £13,000 for different grades of much the same package.

A typical training package comes in two parts: the student unit and the programming unit. The student works with a keypad or keyboard (and perhaps some other control device), a screen and a printer. The video player and computer are also part of the unit, but need not always be seen or touched by the student. The equipment may be used by the programme designer, but usually there is at least one extra piece of kit for the person who is designing and executing the programme. Again, the scope and complexity of these systems varies widely, as do the configurations of hardware employed by each.

Videotape is commonly used in training systems, both for economy and convenience. Of course, no delivery system yet comes complete with a video production studio, so programme designers still have to make the master tape using their own or others' facilities. These delivery systems often cater to organisations already using videotape, that may even have in-house studios. A tape-based system is then self-contained, and people with access to video production facilities can make their own programmes in-house from start to finish, often for little more than the cost of conventional video in a classroom training programme.

* CAVIS (Computer Aided Video Instruction System) was developed in the UK by Scicon, a subsidiary of British Petroleum. It is an expensive, stylish system, tape-based but now tentatively moving into videodisc.

The CAVIS Minstrel system from Scicon, a tape-based PAL-standard package for trainers and students.

* Felix is the product of Felix Learning Systems, who approach the management training market in a more comprehensive way than most other purveyors of new technology. The system is built around U-matic tape and the Apple II microcomputer. There is a range of generic courseware available, and Felix also make programmes for clients—which is a round-about way of saying that customers cannot make their own programmes.

The Felix Management Workstation from Felix Learning Systems, a tape-based, PAL-standard delivery package.

* National Panasonic's Interactive Video Training System can employ ½" or ¾" tape players (the company's own or those of other manufacturers). It is available either complete with hardware, or as a smaller package for people who already have some of the requisite equipment.

* The Sony Responder is a low-budget system that uses a U-matic tape player and a small amount of purpose-built hardware. It is a simple, straightforward package that has enthusiastic adherents despite its limitations. The Responder does well in the UK and Europe; in the US, the combination of Sony's own microcomputer, videodisc player and attendant hardware and software represents a delivery system in itself.

* Philips' 'professional' series can actually incorporate an dedicated 48K CPU into the videodisc player. A range of options including gen-lock and teletext, complemented by Philips' own authoring system, PHILVAS, completes the package.

* Edutronics by McGraw-Hill links either an IBM Personal Computer or a micro from Texas Instruments to a VHS tape player through the CAVRI Interface mentioned in Chapter 21. It employs separate screens for video and computer- generated text and graphics.

* IVIS (Interactive Video Information System) is one of the most forward-looking of the systems currently available. It can incorporate video tape and disc, and has the long-awaited 'graphics overlay' feature for incorporating both video signals and computer-generated text and graphics on an RGB screen at one time. The system was developed by the Digital Equipment Company (DEC) around their own VAX-based computer systems.

The drawback to all delivery systems is their ultimate lack of versatility. The hardware is often dedicated to the sole end of making pre-recorded video tape or disc programmes interactive, so that it cannot be used for any other purpose. (A dedicated microcomputer, for example, cannot do the accounts in the off-hours when it is not running a training programme—which a universal computer can do.)

The authoring system is that supplied with the package as a whole. There is much less opportunity for enhancement, either to upgrade or to expand the system, than there is in configurations built from scratch with compatible multi-purpose computers and video players.

The systems cited above are only a few of the many available, for more are emerging all the time. Choosing a pre-packaged delivery system could be a clean, simple way through the obstacle course—but it could also be an expensive mistake. Again, the only answer is to analyse your own needs very carefully, to shop around, and to ask the advice of someone qualified to comment. Here, as everywhere down the line, our very best advice to you is to make people work to get your business. Don't be dazzled by first impressions or shiny brochures—get a good 'hands on' demonstration, and think seriously about what the equipment under your hands can really do for you.

CHAPTER 28: A HYPOTHETICAL PROJECT

To illustrate some of what we are discussing in this and the next three sections, we have devised a sample project—which is entirely hypothetical. Examples of the kinds of decisions which are made in the course of an interactive video project will be based on this imaginary exercise.

We are taking the perspective of a small production company which specialises in interactive video, and has made programmes on both tape and disc for a variety of clients. We do not have our own video production facilities, although we do have a few different microcomputers. The core of a production team—the people who are involved with a project from start to finish—are on our own staff. People with special skills (camera crew and so forth) are hired as and when necessary.

Our client is the State Museum of Fine Arts—an institution which does not really exist but which, if it did, would be a large gallery holding a national collection of paintings, prints and drawings, sculpture, and latterly, some modern works involving building construction materials and assorted plastic kitchen implements. There is also an collection of small objects—porcelain, silver, clocks and so on.

The Museum has obtained funding for a pilot project to investigate the use of interactive technology for visual information storage, staff training and public education. The project is being funded jointly by the equally mythical Ministry of Arts and Culture and a private foundation.

Proposals were received from a number of companies more or less experienced in interactive video production. The long and nerve-wracking selection process is one stage of the project we will spare you. Suffice to say that on the basis of the proposal and presentation that we delivered, and our track record, we were selected.

In our preliminary analysis, we determined that the Museum has three main aims:

- To catalogue the collections for scholars, students, restorers, lecturers and researchers.
- To train curatorial and other staff.
- To introduce the Museum to visitors and sponsors, and to guide them through its collections.

We will discuss these in more detail in the next section. In the meantime, we have to decide what type of delivery system to recommend, whether tape- or disc-based, and, if disc-based, whether Level 2 (using only the industrial standard disc player's own onboard microprocessor and programs encoded on the disc itself) or Level 3 (employing an external computer).

In fact, this combination of aims points to a Level 3 disc-based delivery system, for several reasons:

* There will be a great number of still frames in the cataloguing alone (there are about 3000 items in the current catalogue). This demands a storage medium which is stable and hard-wearing, offers sharp, clear colour reproductions, is capable of quick and frame-accurate random access, and can hold a still image indefinitely without deterioration.

This points to videodisc–and, specifically, to laser disc–as a medium. Of all the systems currently on the market, laser disc offers the best combination of picture quality, stability, economical still frame storage, rapid random access and special features suited to this project (freeze frame, for instance).

The first generation of VHD players are a close second, falling short principally on still frame storage. Videotape is basically not of sufficient quality or stability for use in this way, even without the problems interactive tape- based systems can have with frame-accurate random access.

The Museum currently uses a combination of photographs and slides as a working catalogue. An interactive slide or tape/slide programme could be based on this. However, the complete catalogue will comprise just over 3,000 images, which (with only one image per object) represents 38 carousels of 80 slides apiece - or a little over 2% of the available 'real estate' on a double-sided laser disc.

The disc is obviously the handier, tidier medium–the whole collection can be held in one hand, the images cannot be lost, mislaid, or put out of order, and they are relatively safe from dirt and damage. (Of course, later acquisitions would be excluded from the first disc. Another disc might be produced in a few years' time; in the meantime, a small collection of slides representing new acquisitions will supplement the catalogue for research purposes.)

Also, while the whole project could conceivably be done as a vast sound/slide show, it will be much more interesting as a combination of still and moving footage. The staff training and public information segments are likely to include quite a variety of shots which will work better moving than still–scenes showing staff at work and visitors touring the collections, for example, and interview sequences. Disc is particularly suited to this combination of material.

The next decision is between a Level 2 and a Level 3 system. Level 2 requires only an industrial standard videodisc player, a screen and a remote control device of some kind; however, the control programs must be encoded on the disc and it would be extremely difficult to add or amend material once the disc is pressed. Level 3–with the addition of an external computer–involves more hardware but is concomitantly more versatile: control programs can be changed or enlarged at any time, new programs added, and volatile information generated and stored on the computer.

Let us consider some more features of this project:

* With some 3000 entries in the catalogue, a detailed information retrieval procedure will be needed both to identify and access individual items in the catalogue, and to caption and/or comment on each entry. A program of this size and complexity must by definition involve an external computer.
* A number of different control programs will be needed to adapt common material (including still frames from the catalogue) to the various training and information programmes.
* A great deal of information in the project will change over time – the arrangement of galleries, for instance, schedules of lectures and public events, guides to special exhibitions, current publications, and even the names of Museum staff.
* It is also foreseeable that changes over the years in things like the administration of the Museum, in the organisation of its collections, and in its marketing and public relations plans, will affect the presentations on the disc.

Considerations of this kind indicate that only a Level 3 system is likely to provide all the features asked of this programme.

In this way – through analyses and decisions – the delivery system is refined to the point of actual hardware specifications. For this project, the initial configuration of equipment will comprise:

* Four purpose-built kiosks in the entrance hall of the Museum, for the use of visitors. These run the public information programme, in a choice of five languages. Visitors are welcomed to the Museum and introduced to its collections, and shown a calendar of exhibitions, lectures and public events, and a catalogue of Museum publications. A videodisc player and microcomputer are housed inside each unit, and a large screen mounted at eye level outside, under a projecting acoustic hood. The light pen with which visitors interact with the programme, and the hand-held audio receivers through which they listen to the soundtrack, are attached by flexible cords to a panel beneath the screen.
* A purpose-built workstation in each of the four main study rooms, for the use of staff and visiting scholars, students and researchers. Each of these has a disc player, high resolution screen, microcomputer, keyboard and printer.
* Two portable workstations, with the same equipment on trolleys which can be moved around the building and even loaned out for exhibitions and special projects.

The longterm plan is to install more units – perhaps in new locations such as main galleries or the lecture theatre – once a pattern of use emerges and as funds become available. Material from the disc catalogue may be sold independently to visitors and educational institutions. It will be easier to determine when and how the project should grow once it has been up and running for a few months, and feedback from its various users has been assessed.

This example, although admittedly simplified, does suggest the sorts of decisions required in a project of this kind. We will return to the State Museum of Fine Arts in the next chapters, using this hypothetical project to illustrate points raised along the way.

CHAPTER 29: SOME WAYS OF PRESENTING INFORMATION

The decision to use interactive video is not so much an answer as the key to a whole new set of questions. Each of interactive video's component technologies – video and computers – has its own strengths and conventions. The possibilities opened up by the combination of the two are so many that people with a background in either one alone often fail to appreciate just how much the two can do together.

You can get a good feel for what video alone can do simply by watching television – and, in the UK particularly, by studying television commercials. TV adverts are currently a minor art form in Britain, fun to watch and extremely inventive. Moreover, the best of them encapsulate, in less than a minute, a great deal of what we are advocating here. They have information to transfer to a selected audience by the most effective means at their disposal within the constraints of the medium they employ and the budget they have been given. The stakes for which they play, at least in material terms, are high.

A good ad agency respects exactly the criteria that the team producing an interactive video programme should:

* Detailed analysis of the subject and its audience.
* Clear, simple objectives.
* Meticulous preparation.
* High production standards.
* Varied selection from a wide range of presentation methods.
* Sensitivity in style and approach.

Advertisements are useful models in that they often employ several different ways of presenting information within a short space of time. Elsewhere during a typical week's schedule, the news, public affairs shows and documentaries, breakfast television, sports coverage, children's hour, game shows, daytime TV, educational programmes and even sit-coms and dramas, all illustrate ways of communicating information to people – information of all kinds, people at every level of the market researcher's scale.

People who normally watch television only during prime viewing hours – evenings and weekends – would do well to have a look in at other times of the day and week. A fortnight's random sampling, with an eye to style of presentation as much as content, is an education in itself. It may even change the way you watch television.

Try to get a feel for things like pacing. How long are you left looking at one thing? How often does the camera move? Are long scenes and short sequences mixed? How does one scene change to another – with an abrupt cut, a smooth fade, or some electronic trick like one part of the screen growing or fading into another? How quickly do people speak?

Consider the texture of the programme. Are you watching live action or material that has been pre-recorded and edited? Is the camera on location, or in a studio? What kinds of backgrounds do you see? How are captions used? How many different sizes and styles of lettering can you identify? When do graphics appear? What kind of information do they convey? How many different kinds of artwork – cartoons, maps, drawings and so on – can you distinguish? Are still frames used as well as moving footage? What kinds of special effects do you see? When do you hear a voice without seeing the speaker on screen – or even thinking about who the speaker is? What kind of background music used, and when? Can you detect sound effects?

Look for style and forms of address. Are you watching 'real people' behaving spontaneously, or professionals at work? How are people dressed? (Does anyone *you* know have hair that looks like that?) What kinds of voices and accents do commentators, newsreaders and presenters have? When are two (or more) voices used in preference to one? When do subject experts, or guest stars, make an appearance? Do you ever recognise a voice, on an advertisement say, without seeing the speaker? How do you feel about the people you see on the screen – do you believe the newsreader, trust the manager of the used car lot, like the stars of the soap opera? Do you think you understand the information being presented? Why do some things interest and impress you, and others leave you bored, indifferent or even angry?

This is just broadcast television we are considering. We haven't touched on the extra dimension of interactivity at all. You begin to appreciate that the man wasn't kidding when he said that the medium *is* the message.

With interactivity, the pacing of the programme can exploit slow motion, step frame and freeze frame. These or the option to use them can be built into specific parts of the programme, or left to the user's discretion. (Remember, it is possible to close off some segments of the programme from the user's control while leaving others open.)

When an external computer is used, interactive video offers even a wider variety of effects than television: computer-generated text and graphics can be mixed spontaneously with video material. Furthermore, the speed at which information is presented, and even the level at which it is pitched, can be tailored to the needs of individual users – a feature many advertisers would dearly love to incorporate into broadcast television.

Before we go on, consider that the British Audio-Visual Association's claim that we receive about 75% of what we learn through sight, 13% through hearing, 6% through touch, and 3% each through taste and smell – but that we remember:

* 10% of what we read
* 20% of what we hear
* 30% of what we see
* 50% of what we see and hear
* 80% of what we say
* 90% of what we say and do at the same time

A good, mixed presentation is more than entertaining – it is effective.

Consider a few of the ways of information can be presented in an interactive video programme:

* **Live action:** a record on film or video of something happening, naturally and spontaneously, in the real world, as it was seen and heard from the perspective of the people with the camera and sound equipment.

* **Re-enactment:** a reconstruction of something that has really happened, possibly using the same people and places, but quite likely edited and polished with the wisdom of hindsight.

* **Dramatisation:** a realistic, but hypothetical, scene which shows something that might have happened, usually with characters and situations equally fictional.

* **Simulation:** a mock-up, using as many genuine props and situations as possible, of something that might happen, usually with the user playing a principal role.

* **Voice over moving footage:** typically, live action with the original soundtrack ('actuality') replaced by spontaneous or scripted commentary delivered by a speaker who may or may not appear before the camera.

* **Voice over stills:** photographs and artwork accompanied by commentary from an off-screen narrator.

* **Captions:** usually, important pieces of information such as names of people, places and things appearing on the screen, but possibly excerpts from commentary (either to reinforce key points or to make more clear words which may be faint or garbled).

* **Text:** a screen devoted simply to words presented on a neutral background, with or without sound. This screen could reinforce key points, or, in the manner of the 'intertitles' of the silent film era, explain or introduce material appearing before or after it.

* **Split-screen techniques:** the screen can be halved or quartered or divided into virtually any shape or proportion, and separate material displayed on each part. The screen might be given over to live action, one detail of which is magnified in a panel along the side. For example, a line drawing of a tool, set against a plain background, could be shown alongside moving footage of someone using that tool.

These are the sorts of things for which you should be looking in television programmes (especially news, public affairs shows and documentaries), in advertisements, in the short features in the cinema, in the films and videos you might see at school or at work—indeed, anywhere you are watching information being transferred.

Of course, this is a huge topic, and one only peripherally related to our subject. The next chapters are designed not to teach the average reader how to become a television producer, but simply to give you a feel for the kinds of ideas which go into the design of a programme, whether an interactive instruction manual or a TV commercial.

CHAPTER 30: MOVING FOOTAGE, STILLS AND SCREEN DESIGN

In the last chapter we suggested the great variety of visual material which can be incorporated into an interactive video programme. Now, let us consider in more detail different kinds of still and moving footage, artwork and graphics, and the strictures of design for the small screen.

REAL ESTATE

Space on a video tape or disc is sometimes called 'real estate'. This is a good image, for it is useful to think of interactive video in terms of space as well as time. The linear playing time of an interactive tape or disc is no reflection of the time any one user may spend with the programme. Since some ways of conveying information use less space than others, design is a material as well as a stylistic question.

Consider the humble still photograph. People tend to think of video primarily as moving footage, which was natural enough when tape was its sole medium. As we have seen, so long as there is contact between the reading head and the fragile surface of the tape, videotape cannot hold a still frame for long without suffering wear. For that reason, tape tends to record photographs and artwork on footage moving at 25 (PAL) or 30 (NTSC) frames a second. So stills traditionally have had little material advantage over live action, whatever the stylistic considerations. (The latest tape players offer many impressive features, but the problem of wear on the tape is still an obstacle to the use of stills and freeze frames in a tape-based system.)

Laser disc, on the other hand, can hold any frame indefinitely. This could be either a still prepared as a piece of artwork (a photograph or drawing, perhaps) or a freeze frame taken from a piece of moving footage. As we saw in Chapter 15, the two capacitance systems, VHD and CED, are not yet quite so versatile; however, both can still hold one frame for minutes at a time.

Furthermore, in any disc system, a still frame, no matter how long or how often it appears on the screen, occupies at the worst (in a first generation CED player) twelve frames. On a laser disc, a still takes up only one frame of a possible 54,000 per side. This is a far cry from tape's 25 or 30 frames a second, that squanders 250 or 300 frames on a ten second still. With the introduction of videodiscs, the still frame suddenly becomes a highly economical means of information transfer. Alternatively, if information can be conveyed through a page of computer-generated text or graphics, these can be substituted for still frames recorded on the disc or tape.

So, when the delivery system employs videodisc and/or an external computer, it may be worth using a still frame or a page of computer-generated text or graphics in place of moving footage.

Let us then consider the many types of still and moving pictures that can be used in interactive video.

STILL PHOTOGRAPHS

Still photographs can be as effective as moving footage to illustrate hard information. A photograph is often easier to study than moving footage, especially if the user is looking for detail. And don't forget that a series of stills can be used in place of moving footage to equal effect and often more economically. Even an action sequence (the demonstration of a manual skill, for example) can be conveyed by step framing through a set of carefully graduated still frames.

Still photos are less satisfactory when addressing 'soft' skills or 'people' skills, although a set of stills can be used to supplement a sequence of moving footage once the mood has been established. For example, the feeling of, say, an interview can be established on moving footage in a dramatisation with professional performers in both roles. Either part can then be handed over to the user, who can pose questions or give answers interacting with a series of captioned stills.

FREEZE FRAMES

Freezing on a single frame of moving footage is an effective documentary technique. It is a dramatic way to illustrate an important detail or end a scene (if you saw the film 'Butch Cassidy and the Sundance Kid', you undoubtedly remember the last shot – a superb use of this effect).

A single clear frame from a segment of moving footage can also be used as a still. As well as being economical, this, too, has the effect of reinforcing the visual image on the user's mind. As well, allowing the user to freeze on any frame of a segment, and step frame through a sequence which can also be seen as moving footage, can be extremely useful in explaining, for example, the intricacies of a manual or mechanical operation or event – a technique in surgical stitching, for example, or the subtleties of wave motion.

ARTWORK

Artwork is the term used in publishing and production to describe illustrative material. It is a generality which for our purposes we will restrict to mean work actually prepared in traditional media (such as paint and ink, paper and card) by commercial artists, or archival material unearthed by a researcher and prepared for the screen by a screen designer.

Consider what this can include:

* Reproductions of fine art – paintings, prints and drawings of all kinds.
* Popular art, including posters and handbills.
* Illustrations of bygone days and ways.
* Reconstructions or 'artist's conceptions' of scenes, places or objects.
* Sketches, cartoons and caricatures.
* Maps old and new, political and geographic.
* Blueprints and technical diagrams.
* Clippings from newspapers and other publications.
* Facsimiles of historical documents.
* Samples of handwriting and manuscripts.

Some of these could of course be produced as computer-generated graphics; that, as a separate topic, is discussed in Chapter 23.

Artwork gives programme makers more control over what the user sees than do photographs. A drawing is often better than a photo for conveying impressions unqualified by superfluous detail, or drawing attention to some feature or activity which might escape the naked or untrained eye. Building up a series of graduated sketches is a good technique for introducing complex information gradually.

The creative use of photographs and artwork of various kinds and vintages can also improve the texture of the programme. Archival material, from picture libraries or other sources, often adds interest and even a touch of light relief.

CHARTS AND GRAPHS

Graphic representations of financial and statistical data are a category unto themselves, and a problem area, too. Many people simply turn off at the sight of a chart, because bad graphics have made them believe that they will never understand information presented in such a way. This is sad, because much valuable information is conveyed to the cognoscenti through charts and graphs, to the exclusion of people confused by unintelligible graphic language.

Unfortunately, there is as yet no standard graphic language, so every chart and graph must be taken on its own terms, and deciphered – or not – from the directives which accompany it. Neither is there a standard colour code, even though many colours have strong emotive values, and certain professions have their own conventions (accountants' use of black and red is one well-known example of this).

Typical charts include:

* Bar charts, vertical and horizontal.
* Variance charts.
* Component charts.
* Step charts.
* Pie charts.
* Scatter diagram.

The superficial form of these various types of charts is illustrated below.

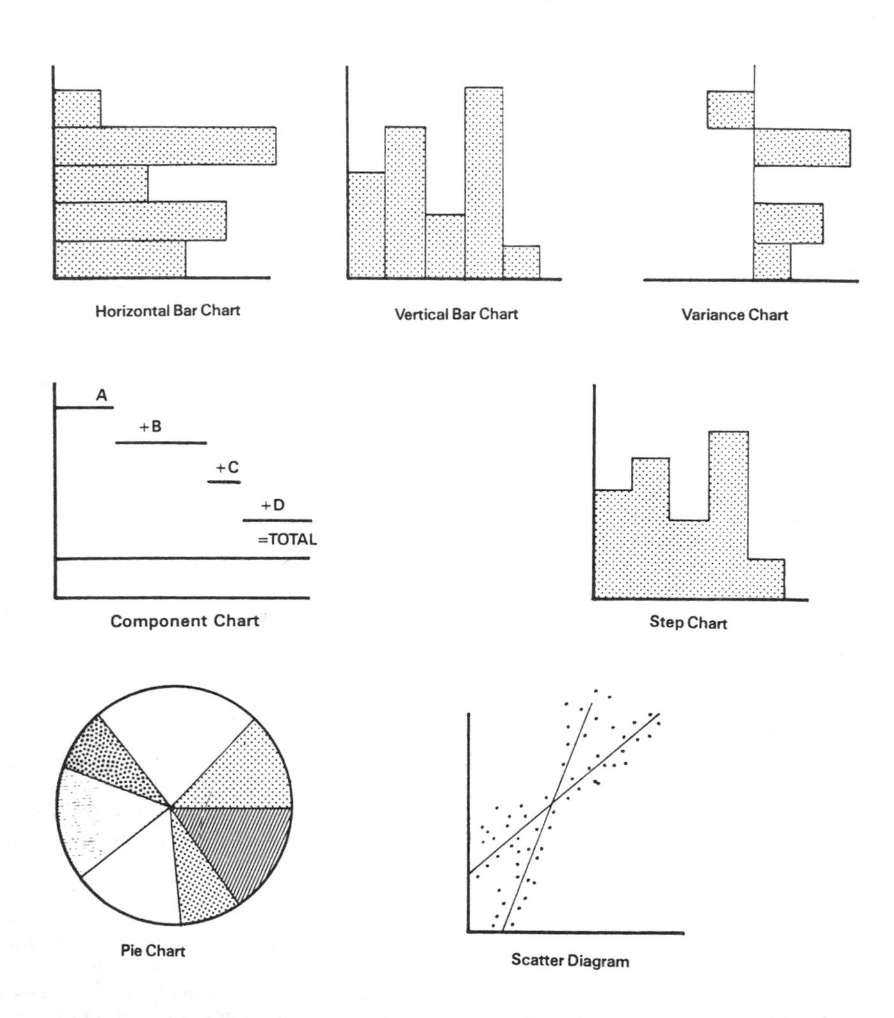

Horizontal Bar Chart

Vertical Bar Chart

Variance Chart

Component Chart

Step Chart

Pie Chart

Scatter Diagram

There is a distinction in the computer industry between 'presentation graphics' and 'information graphics'. They may sometimes look alike, but their purpose is different.

* Presentation graphics convey in graphic form information already apprehended in some other way, through study, discussion or explanation. They drive home points which have been approached from other directions.

* Information graphics represent information which cannot effectively be conveyed in any other way–particularly that which establishes a relationship between different pieces of data.

Whether charts and graphs are likely to be effective depends very much on their being used appropriately, designed intelligently, and addressed to a receptive audience. A group of bankers would probably feel perfectly comfortable with the same charts to which passers-by in a suburban shopping centre could be quite hostile.

Generators such as the Dubner and Chyron IV cited in Chapter 23 are machines which help design and execute text and graphics for the screen, and one of the best means of creating clear graphs and charts. Most video facilities houses have a graphics generator of some sort, to provide 'supers' (captions superimposed on the video picture) if nothing else. Many have top-flight equipment, and those that do should also have demonstration tapes to prove it. (If you do decide to use a powerful graphics generator, make sure that you also employ a qualified operator.)

The greatest error in screen design, and one commonly made, is trying to force too much information into the space. This is particularly a problem with graphs and charts, and is only aggravated by the strictures of the small screen. A few tips on screen design are included at the end of this chapter.

ANIMATION

At the opposite end of the scale from humble pie charts are animation sequences. If you tried the television watching exercise outlined in the last chapter, and paid particular attention to the advertisements, you will appreciate that there are a great many more kinds of animation than the sort represented by Mickey Mouse and Bugs Bunny.

If you live in or near a cosmopolitan centre, you might be able to attend an animation festival, or see an anthology of animated short features at a film society. This would give you a feel for the wide range of techniques available - and for the cost of producing animation. Walt Disney studios in their heyday produced hundreds of thousands of drawings for their spectacular animated features. Compare the quality of a film like 'Fantasia' with modern Saturday morning children's programmes–which are still quite expensive to produce–and you will appreciate how much time, effort and money goes into even sixty seconds' worth of high quality animation.

New techniques employ a variety of media from plasticine to finger-painting–and, of course, computer graphics. A spot of simple animation could well be within the budget of even a fairly small project, but the decision to go for what is usually a luxury must be made carefully. Animation in the form of computer graphics has been used impressively in, for example, a videodisc on anatomy which illustrates the workings of the brain and the inner ear–neither of which is really a good place for a location shoot.

More modestly, in a project of any kind, one brief animated segment, since it can be used again and again, is often a good investment. A cartoon character or a motif such as an animation of the organisation's logo or symbol can be used to effect throughout the programme. For instance, an cartoon character could:

* Welcome users to the programme and guide them through it.
* Link disparate segments or introduce new parts of the programme so that the user does not notice a sharp change.
* Add a light touch to a long or dry programme.

Poor use of animation can be as embarrassing as the poor use of humour—especially when the two amount to the same thing. Taste, tact and talent are required to control the tone of a programme in this way.

SCREEN DESIGN

Here we have yet another subject which is a career in itself. Few people realise how specialised a discipline screen design is, for there are many talented commercial artists who have no idea how to work with a video or computer screen. A great deal of time and money is wasted in facilities houses by people who turn up with artwork to be put onto film or video, only to discover that what they have simply won't work on a screen.

The ideal solution is of course to hire a qualified screen designer—that is, someone with experience in television screen or computer graphics design (remember, even those two are not the same). Alternatively, consult the people who will be doing the transfer of conventional artwork to slides, video or film. Most facilities houses are only too happy to lay down guidelines—and avoid trouble later.

For instance, consider that a screen has a fixed 'aspect ratio'—that is, the ratio between the width of the screen and its height. The aspect ratio of a TV screen is four units of width to three of height. In film, the ratio is three of width to two of height.

This imposes strictures on the composition of any picture, whether drawn on paper or shot by a video camera. For one thing, in professional terms, it means that all artwork for the screen must be prepared in 'landscape' rather than 'portrait' format. If you want to show a tall building, you will have to show a lot of the land to either side or, as one enterprising advertiser did, tilt the picture on its side to emphasise how tall the building really is.

Furthermore, the screen tends to cut off the edges of the picture—ten per cent of the total area of the screen, around its perimeter, should be left clear of information. The area which is clear of this loss is called the 'safe area', and all information should stay within its bounds.

You will probably now notice that much of the artwork you see on the TV has an identifiable safe area. Many photographic cameras have the same safe area outlined in the viewfinder—the principle is the same.

Then, certain colours simply do not work well on a video or computer screen. Strong colours, particularly reds and oranges, tend to 'bleed' into the screen. These are called 'saturated' colours, and they should be avoided as much as possible. If, of course, you are making a video for a company whose logo is scarlet, you simply have to do the best you can.

Also, because the video screen is actually made up of hundreds of fine lines (the scan lines explained in Chapter 10), some ghastly effects can be produced if, by chance, artwork is produced on a background of very fine lines. The chances of this may seem slim, but it has happened–and the artwork had to be redone. The story is cited principally as an example of how many weird and wonderful things can strike you out of the blue if you do not prepare, check and double check at every stage of the production.

These are technical considerations. Let us also think of a few stylistic problems. As we said, the commonest error is trying to fill the screen with information. This is easy to do, and very hard on the end user. It is totally off-putting to be confronted with a screen jam-packed full of text or charts, and difficult to decipher such information. This is one place where less really is more:

* Use text sparingly. Space sentences, paragraphs and key points so that the user can distinguish each separate idea at a glance. One point per screen is a good ratio.

* Use bullet points, indentation, numbers or letters, colour or any other layout device to identify individual points and draw attention to important information.

* Since you have the benefit of animation, use a flashing cursor or some other attention-grabbing device to direct the user's attention.

* Instead of bringing a page to the screen full of information, in one complete block, try building on an empty screen. Many authoring systems allow you to lay down 'running' text, which prints out a character or a line at a time, as though it were coming off a printer. You can even control the speed of this (250 words a minute is an average adult reading speed). Similarly, you can build up a chart or a diagram one element at a time, so users can follow the logic behind it.

* In general, follow the journalists' rule of thumb–'when in doubt, leave it out'. They, of course, mean leave it out altogether; we suggest that you look at every busy screen as the penultimate design for two separate screens. No matter what your presentation, it rarely hurts to break information down into its simplest components.

* Here, more than anywhere else in the programme, choose your words carefully. Again, the fundamental rules of popular journalism are a good model. Aim for simple, declarative sentences and verbs in the present tense. Avoid the use of negatives. Choose short, unambiguous words in preference to pompous or unusual ones. Only introduce jargon or technical terms when these are a part of the information being conveyed; even then, make sure that the meaning of new words is clearly conveyed, either explicitly or in context.

This is really only the beginning, but it has served its purpose if it has alerted you to some of the rules of the discipline–or even made you aware that there is one.

THE MUSEUM PROJECT

Finally, let us return to the hypothetical project we set up in Chapter 28, and illustrate this discussion by considering some of the material we are putting into our programmes for the State Museum of Fine Arts.

* Naturally, since one objective of the programme is to catalogue to Museum's collections, a part of the disc will be devoted to still frame storage of roughly 3000 reproductions of paintings, prints and drawings, photographs of sculpture and so forth. The Museum already has colour transparencies of a many of these, but many more will have to be photographed specially.

* Live action footage will be used for 'establishing shots' (showing the audience where we are—whether the entrance hall or the Post-Impressionists Gallery, for instance) and for 'beauty shots' (interesting or attractive shots which can be used as background to narration or as a filler between segments). This will be taken in and around the Museum.

* Archival footage from the War Memorial Institute will be used to show how the Museum coped during the Second World War, removing and protecting its collections, and dealing with bomb damage. This footage is in black and white, on cine film, and in rather poor condition.

* We are also using an excerpt from an American broadcast television programme which covered a touring loan exhibition, to show how the Museum's collections sometimes travel long distances to appreciative audiences.

* There will also be some specially-commissioned artwork. That used in the staff training portion of the programme is basic and unlikely to change, and will therefore be recorded along with the video material in the final edit. Things like the arrangement of galleries, for instance, are likely to change over the years, so that type of information will be put on the computer, so that it can be updated.

There will be some more live footage, but we will wait until the end of the next chapter to introduce it.

CHAPTER 31: PEOPLE, VOICES AND FACES

There are many places for people to appear in a programme, and many roles for them to play. Think of some of the ways people can be employed:

* If you want to demonstrate a product, a machine or an activity–whether for a point-of-sale unit or a training project–don't forget to show people using the object or doing the job. Users will pick up far more than you could ever think to tell them simply by seeing people like themselves, demonstrating a skill.

* If you have a lot of dry, technical information to impart, try to relate it in some way to human events. An interactive videodisc on physics, called 'The Tacoma Narrows Bridge Collapse' is a classic example of this. It bases a lesson on wave motion around historic film footage of a suspension bridge which broke apart during unusually high winds (fortunately, in front of someone with a cine camera).

* If the programme involves some form of problem solving, try to illustrate the problem from the perspective of people who face it. Get a series of short interviews with real people saying what they think. These are called 'vox-pops' in the trade: you often see them used in advertisements. (If in the beginning of the programme, you show people saying what's wrong, do by the end of it have them comment on what's been put right.)

REAL PEOPLE AND OTHERS

Never forget that you can use both 'real people' and 'professionals'. Sometimes, the idea of using one or the other, or both, does not occur to programme designers.

Professionals include TV personalities and other well-known faces and voices, subject experts, and paid performers–particularly, people acting out scenes which are not typical of their real lives. This is a question of style and tone. Sometimes a star turn adds interest and polish to a programme–and sometimes the sight of actors and actresses playing at jobs of which they obviously have no practical experience can seriously undermine the credibility of a presentation.

On the one hand, you might be surprised to know how easy it is to hire the services of the familiar and the famous. If you were listening carefully to the voices during the television-watching exercise outlined in Chapter 29, you should have caught the voices of quite a few popular stage and screen stars during the commercials. If your organisation uses corporate video, you may well already have seen national or local TV journalists presenting your company's business. These people often don't charge more (or much more) than unknowns – which is one reason why they're so familiar. And, with video, you can afford to use top talent: the performance that is recorded once can be seen times without number, far away from the time and place in which it happened live.

On the other hand, you might find it surprising to learn that in many situations, people would rather see an unpolished performance from their peers and colleagues than an artifical one from someone who is obviously not one of their number. And you might be surprised to learn how many little details will give away the poseur. Real people, in situations familiar to them, sound and look natural. You can sometimes save yourself a great deal of work and frustration simply by using them.

A PRESENTER AND/OR NARRATOR

You might consider employing an identifiable figure or voice to introduce the programme and guide the user through it. A presenter pulls a programme together, and provides a handy link between segments. This person can act simply as the narrator, or can appear on screen sometimes too. The use of an on-screen presenter can lend a more human touch than that of a voice from a narrator who is never seen.

You have an interesting choice here:

* A well-known TV presenter, newsreader or anchor – someone associated with broadcast journalism – lends authority to the programme, and will be adept at television presentation techniques.

* Similarly, a celebrity from some other field can often add weight or dimension to the programme, as well as an element of glamour and surprise. Audiences are sometimes flattered that a famous face or voice is showing an interest in them and their world.

* Then, too, there may be someone within your organisation with the requisite broadcasting skills, who will appear sufficiently relaxed in front of the camera and be familiar to your audience. Don't discount this possibility – the use of 'real people' always lends integrity and realism to a programme.

Again, it is important to choose people likely to appeal to your audience. It is counter-productive to employ a presenter whom the audience feel is 'plastic', uninterested in them or the programme, or of the wrong calibre for the information being discussed. The presenter must engage and sustain the users' attention, trust and co-operation.

VOICE-OVERS

The disembodied voice you hear talking over a scene in which you do not see a speaker is called a 'voice-over' (often abbreviated as 'V/O' or 'v.o.'). A voice-over is often a practical alternative to quantities of text, and a welcome change from the sight of talking heads.

Remember, mixing methods of presentation improves not only the flow of the programme, and the user's comprehension of the material being presented, but also the likelihood of the user's remembering that information. The combination of a voice-over and actuality enhances the impact of both. The reinforcement of key words or ideas through captions or bullet points on the screen, and even by a quick test, makes for a highly effective presentation.

A voice-over can accompany virtually any visual material – actuality (live action shot on location), archival material, a dramatisation, photographs, still or animated graphics, diagrams, charts, maps – all kinds of things. The visual material may either illustrate what the voice is saying, or, alternatively, convey information which benefits from intelligent commentary.

The choice of voice(s) is important:

* It may be the voice of the presenter or anchor, if there is one, or some other key figure in the programme (such as a subject expert) which is used.

* An attractive documentary technique, especially good with material collected in interviews and other unscripted conversations, is to run the voices of 'real people' over actuality (this is called 'wild track'). The footage should reflect the content of the voice-over in some way.

* Sometimes it is useful to employ special voices just for voice-overs. This tends to distinguish, and sometimes to distance, voice-over sequences from other parts of the programme, which can be desirable.

* If you are using separate audio tracks to bring out different aspects of the same visual material, it is a good idea to use different voices for each - one voice for the sales pitch to customers and another for the sales staff training, for example.

Never forget that voices have potent associations: different timbres and accents are perceived as appropriate to different kinds of information, and to different audiences. An affected or unfamiliar accent can be as off- putting as a shrill tone or garbled delivery. These are critical points of style which can put users at their ease, or irritate them irremediably.

THE MUSEUM PROJECT

Now, by way of example, let us return to our hypothetical project for the State Museum of Fine Arts, introduced in Chapter 28. The Museum and its collections will be introduced in both the public information programme and in the first of the staff training segments by a professional presenter, who will appear on screen and do voice-overs in a brief guided tour of the building and important works in its collections.

For this job, we have chosen a presenter who has made a TV series introducing art history, and who is a popular figure on radio and television. This person will be familiar to many visitors, and has the style and confidence to appeal to those who don't recognise the face. With a background in art history, this presenter, too, is at ease with the material and in the environment.

For the staff training sequence, one of the museum's own lecturers, who is a fluent and entertaining speaker with a good deal of specialist knowledge to boot, will be reading the voice-overs and helping prepare the script. Museum staff will recognise this figure, and feel comfortable under the tuition of someone they know and respect.

We will be interviewing some important people in the world of museums and art history. We debated interviewing them all separately, in their various offices and homes, but eventually decided the trouble and expense would be too great in proportion to the contribution this would make to the programme–so we compromised. We will interview the Director of the Museum in the Boardroom and some of the Keepers in the galleries and workshops connected with their collections. The guests not directly connected with the Museum will all come to a studio for a one-day shoot.

Finally, there will be a series of vox-pops with visitors to the Museum, both tourists and scholars, and with members of the staff, both curatorial and commercial. Tourists will be asked whether they enjoyed their visit, researchers will explain why they come to the Museum, and various members of staff will talk briefly about their work.

As a finishing touch, we are planning two establishing shots with which to begin and end every version of the programme. Everyone who uses the disc–scholars, visitors, staff–will enter through a panoramic shot of the Museum building in early morning light, and leave on a close-up of the Museum's cat snoozing in a patch of sunlight, curled up in a favourite nook on a piece of heroic sculpture.

CHAPTER 32: SOUND EFFECTS AND MUSIC

The use of actuality sound recorded on location, sound effects and music must not be overlooked. Used judiciously, sound and music can convey important information, enhance the presentation as a whole, and work subtly on the user's response and comprehension.

People 'watching' TV or movies are only occasionally aware of hearing background music or sound. Yet they are often manipulated by it, and would notice its absence, even if it were not immediately clear to them what was missing. Remember the great cliché of suspense drama: "It's *too* quiet around here, Van Helsing." Any scene which involves lots of people or machinery–a street scene or a factory shopfloor, for instance–becomes surreal without background noise. Even if actuality runs quietly beneath a voice-over, it still adds critically to the effect of the whole.

Sound effects ('FX' in the jargon) can be used realistically to add atmosphere - and also humorously, to set off something that has just been said or seen. You can record sound effects on location or use standard sound effects from the collections for sale or hire through music libraries. A typical catalogue will offer a tremendous range of effects, from authentic bird calls to blood-curdling screams. (Well, you never know what might come in handy.)

Your hours in critical appraisal of broadcast television will prove to you what a potent influence music is on our interpretation of information. Often, you wouldn't know whether scenes in films or TV shows were intended to be peaceful or sinister, serious or satirical, incidental or important, without the music. Think about this the next time you watch some innocuous scene, like the interior of an empty room, or a landscape, and hear the music 'telling' you what you are supposed to feel about it.

Music has a role in corporate video, too. It sets the mood, as it does in drama, whether as the theme for a programme or in the transition between scenes. Music is relaxing, and a brief scene with music over 'beauty shots' makes a welcome break between scenes which may be difficult or tiring for the user. In many training programmes, it can be useful to put instructions on one audio track and music on the other, so the user can practice with visual guidance alone. Many of the 'teach-yourself' exercise discs in the consumer catalogue offer this feature.

You can commission special music for your programme. You can also use familiar classical or popular music, never forgetting that most recordings are subject to copyright. Rarely will the holder of the copyright object to your using the music, only to your using it without having paid for the privilege. Sorting out copyright is one job you can be happy to pay a production company to handle - it's fairly routine for someone who knows the ropes, and a real headache for anyone who doesn't.

Popular music is often expensive, because its market value is high but short-lived. You will find that most production companies rely on 'standard music' libraries. These produce albums by the hundred-weight, all of 'general-purpose' music which anyone can use for a fee. The music is of all kinds from pseudo-classical to electronic, and covers just about any mood or situation. The titles, which are often pretty hilarious, are (unlike a lot of song titles in the top forty) meant to tell you about the music: 'Dramatic Impact (6)—Shock horror disclosure', or 'Press Release—Positive theme with a nice sense of urgency'. If you learn to recognise any of these tunes, you may be surprised to hear where they turn up—they are cheap and handy, and everyone in the industry uses them.

You might also consider using archival sound as well as archival footage, to add interest and dimension. Sound archives and music libraries often have interesting collections recording great moments and voices in history, popular entertainment of the past, and special effects. Music libraries and sound archives are listed in video industry directories and in the yellow pages. Most production companies have their catalogues and a good collection of their albums and tapes—and you are usually welcome just to go along and poke around.

One cautionary note: it is safe to say that, unless sound or music is the subject of the scene, the effect is usually overdone if the user is distracted by it. This is a warning, for sound is 'fun', and it is only too easy to go overboard on special effects and music accompaniment.

In the Museum project, period music from the fifteenth century onwards will be used to accompany beauty shots of the various collections, concluding with a medley of genuine top ten tunes as we survey the Pop Art collection. We were lucky to find a rare wax cylinder recording of an elderly French Impressionist describing his garden, which we will use briefly when we come to that period in art history. For the opening, closing and linking sequences, we have selected a theme from a music library, a lively little neo-classical number which we feel conveys the spirit of the project.

CHAPTER 33: OBJECTIVES AND EVALUATION CRITERIA

Since the dawn of human enterprise, the people who actually make and do things have had difficulty extracting detailed briefs from the people who give out the commissions. People generally have a tendency to get excited about the prospect of getting or doing something without any very clear idea of what they really want. The architect Imhotep probably had to pull teeth to get a specification for the Step Pyramid out of King Djoser. Quite recently, an excellent report on interactive video was presented to people who thought they had commissioned a study on interactive television.

The importance of good planning cannot be emphasised enough. Technical considerations are one thing: you can spend a lot of time and money on a video programme that just doesn't look good. But there are other, more subtle factors which profoundly effect the success of the project. You can make a glamorous programme that is impressive technically, and still fail utterly to accomplish what you set out to do.

Discovering what you really mean to do is a job in itself. This is the first stage of the production process, and you should not be surprised if it seems to take a long time – or even if the job of the consultant begins here, helping work out the real aims and objectives of the project.

An idea that seems perfectly clear in your mind becomes strangely elusive when you try to express it on paper. You may find that your impressions change critically in the course of being transferred from abstract thought to precise language. A brief is almost always revised once it has been shuttled between the client and the production company – or their equivalents – a few times. (If it doesn't, either the original document was a thing of rare truth and beauty, or no one really understands what anyone else is planning to do.)

The distinction between aims and objectives is deliberate – a reflection of how this process of honing and refining basic ideas often works. Asked for objectives, most people usually come up with aims – broad, encompassing, abstract. 'We want to make a beautiful programme that everyone will enjoy.' Aims are expressed mostly in subjective terms, but objectives need clear, concise language. 'We want machine operators to learn to use the equipment that we are going to introduce next summer.'

Determining the aims and objectives of the project is the foundation of all that follows – the instructional design, the flowchart, the video programme, the works. The stated objectives of the programme are the basis of many subsequent decisions, creative and technical: if the objectives are wrong, the decisions are likely to be wrong, too.

Furthermore, the success of the finished programme depends on its achieving its stated objectives. If it has some other, unanticipated, happy effect, all well and good. But it is not in the finest tradition of instructional design to make a programme and then see what it does, although in practice this often happens. The point of good planning is to obviate surprises, even pleasant ones.

A key figure in the interactive video production team is the Problems Analyst, who identifies problems and proposes viable solutions. Before you commission a programme or engage a production company, you need to become a Problems Analyst yourself, and the first thing you need to analyse is whether you have a problem worth solving. This may sound absurd, but it is an approach which will help marshal your first thoughts in some practical way. Here as everywhere else down the line, you may find that, after analysis, what you have is not what you thought you had. The problem may not be so great as you first imagined, it may even evaporate; however, given the way of the world, it may well be larger, different, or more complex than you thought.

Aims may start in the realm of the abstract, but they should come down to the ground with a bump, very quickly. If the objectives are to be the basis of something as concrete as a flowchart, and represent the criteria against which the ultimate success of the programme is measured, then they, too, must be concrete.

Abstract verbs – like 'understand' and 'appreciate' – may help clarify your thinking, but they are not objectives. Trying to determine whether the student 'knows' the lesson can lead to any number of subjective evaluations. It would be more to the point to determine, for example, whether the trainee maintenance engineer who didn't know the difference between an Allen key and a screwdriver when the lesson began can now distinguish between a Phillips and a Pozidriv.

Terse, descriptive verbs – words like 'write', 'repair', 'solve' – describe objectives. Objectives should describe what you expect the user to do, or be able to do, at the end of the programme – whether to buy something you are trying to sell, or to fix something you want to have repaired. Once the objectives are expressed in precise, active terms, the success of the programme can be measured in the user's ability to meet those objectives.

Setting down a list of objectives can be a useful exercise in many situations. A casual list jotted down on the back of an envelope can help clarify ideas thrown about in a brainstorming session. A formal list, clearly expressed and ranked in order of priority, obliges everyone involved in the project to think about it in the same orderly way. A crisp, unambiguous list of concrete objectives, understood and agreed by everyone at the discussion table, is one of the most valuable documents in the project.

Furthermore, the objectives are also the foundation of the evaluation criteria, which are vital to any project – particularly a pilot which may lead to more work on a larger scale. Evaluation represents a way of appraising the worth and effectiveness of the programme and identifying possible improvements which could be incorporated into later work. The value of any project, of course, may be different to different people – management and workers may not see the same programme in the same light. Any formal evaluation structure must respect this potential conflict.

Evaluation begins during the production stage, becoming more complete and more formal when the project is implemented and feedback can be monitored realistically. Ultimately, the evaluation sets out to:

* Assess whether the programme has met its stated objectives.

* Provide constructive criticism that may help improve subsequent work.

* Determine whether the project should be developed on another scale and, if so, in what way.

The evaluation may be conducted in a number of ways–by a questionnaire, in discussion, through interviews, with the computer's own record-keeping facilities or through a combination of these. Undoubtedly there should be a formal document setting out the evaluation criteria and procedure. The principal points involved include:

* Setting out the stated aims and objectives of the project.

* Identifying any other factors that may need to be evaluated (such as maintenance and running costs).

* Identifying the different interest groups whose needs and expectations must be considered, and the criteria for evaluation within each group.

* Determining the best means of evaluating each of the stated objectives.

* Designing the physical means by which data will be collected, stored, processed and output.

* Setting a reasonable time-scale for the evaluation of each point.

* Agreeing who should be responsible for evaluation of each of the objectives.

The same principles which apply to setting objectives are applicable here. Ideally, the evaluation may be summarised in point form on a chart of a few pages. The evaluation may be a single finite job leading to a report and discussion paper, or may be a continual process.

Let us return to our hypothetical project for the State Museum of Fine Arts. Broadly, we have defined the aims of this project as:

* Cataloguing the existing collection on videodisc.

* Providing a public information programme for visitors to the Museum.

* Creating an induction programme for Museum staff.

Only the first of these represents a practical objective: it is quite clear what is meant by 'cataloguing the existing collection', and both the work involved and the criteria by which that work can be evaluated are implicit in that statement.

The Museum, of course, catalogues every object that comes into its possession. Our brief is to transfer at least one complete photographic reproduction of every item in the current catalogue onto the disc, and to assign a unique reference number to each frame. There are roughly 3,000 items in the present catalogue, some of which will take up more than one frame, since details will be shown of large or intricate items, and some objects will be shown from more than one angle. Organising all this will be a considerable administrative job, but not a great creative challenge.

The other two points will need to be refined before they represent concrete objectives. We might make our aims more specific by deciding that we want to:

* Welcome visitors to the Museum, and introduce them to its collections.

* Alert visitors to the exhibitions, lectures and special events in the Museum's calendar and encourage them to take advantage of these.

The first of these reflects the promotion of the Museum as a whole and the second, that of its public education services. The success of this part of the programme will be difficult to analyse, for visitors to the Museum are guests and cannot be tested or quizzed in the same way as, say, employees, to determine whether they have benefitted from what they have seen. The visitor who chooses the 'Introduction to 20th Century Painting' can hardly be hauled aside and obliged to name three Fauvists before being allowed into the gallery.

On the other hand, at least an indication of the success of the project may be measured in improved attendance figures. Since the Museum already keeps figures both for the institution as a whole, and for selected galleries and events, the list of objectives could include:

* To increase attendance at (a) special exhibitions, (b) public lectures and gallery talks, (c) concerts and film shows in the Lecture Theatre.

* To attract more visitors to the collections in the side and upper galleries, which are less accessible than the large central galleries and house collections with less popular appeal.

As to assessing the appeal of the broader part of the programme, we suggest using the record-keeping capabilities of the system to maintain a running survey of visitors' impressions and requests. When visitors opt to end the presentation, they will be asked to answer a few questions before leaving, using the light pen to indicate 'yes or no' and multiple choice answers on the screen. The computers keep both cumulative results of this questionnaire and of the pattern of use of the programme as a whole. This will indicate which segments are most popular and successful, and which either do not appeal or succeed as well as they might. Every week this is summarised on a single-page printout which can be used in planning new exhibitions, facilities and events.

The staff training portion of the project will be more complex. However, it represents an exercise based on principles laid down when Plato was a lad and models from the relatively well-worn ground of computer based training (CBT). The course designed for this project is essentially an induction programme, although it is likely that more specific topics will be addressed in subsequent discs if this one goes down well.

We won't go deeply into this part of the programme here, except to offer a few random examples of the kinds of objectives which will be set for different segments of the programme. These include points like:

* Familiarise all staff with the fundamental information about the Museum: the number of items in its catalogue, the cultures from which it has material, its most famous and valuable pieces, the number, size and function of its internal departments, and so on.

* Teach Grade 5 curators how to pack both flat and irregularly-shaped objects for moving, storage and shipping.

* Teach Grade 3 curators how to arrange display cases: how to open and shut them, how to fit shelves, fix hanging objects, attach labels, connect lights, and so forth.

* Demonstrate the operation of equipment in the Lecture Theatre to Education Department staff.

The whole list, obviously, would be quite long – this extract simply conveys the idea. Although there will be many paths through the induction programme, some portions will be seen by everyone – everyone, for instance, will have to know that the Museum's prize piece is 'The Laughing Fishmonger', and be able to name the 17th century Dutch master who painted it.

Furthermore, Grade 5 curators will have to be able to identify the materials needed to pack a framed oil painting on canvas for a journey by train. Grade 3 curators will have to list, in order, the tasks involved in lining a display case with baize. Such skills need to be practised with real materials, of course, but with a choice of four permanent workstations and two portable ones, this should be possible to arrange in one way or another.

It should be clear by now why the setting of objectives is important – and why this stage of the project can take a long time to complete satisfactorily. It should also be clear that, by the time a comprehensive list of practical objectives has been agreed, a great deal of the information needed in the next stages of the project has been assembled, and a good deal of the thinking has been done, too.

In the production schedule, the setting of objectives runs roughly alongside preliminary research, and the two often lead to the same end. It is important to remember that false economies now can amount to expensive mistakes later. This is as true in the early, intellectual stages of the project as it is in the later, mechanical ones. It is as important to produce a clear, comprehensive brief as it is to shoot good quality video footage. It was clear enough to Imhotep when he designed the Step Pyramid – it should be clear to the production team, too: what is built on a sound foundation will last.

CHAPTER 34: AUDIENCE ANALYSIS

The very phrase 'audience analysis' conjures up the worst excesses of the broadcast television or advertising industries, and the meticulous market profiles for which they are sometimes satirised. But defining the audience is at least as important to an interactive video programme as to the sales launch of a new show or product. The analysis can even be carried out in the same way, although it is to be hoped that in the interactive video project, it will lead to more profound conclusions than those often reached by the creators of TV sit-coms or the people who think up things to put in aerosol cans.

Defining the audience is a task akin to setting objectives (the topic addressed in the last chapter), for it usually involves some basic research, and a good hard think about what the project is really aiming to do. The same tenets also apply—setting down a list of potential users, in some order of priority, often both helps to clarify the scope of the project, and reveals that the aims of the initial brief are too many to be met satisfactorily by one programme.

Sometimes, then, audience analysis falls into two parts: defining the potential audience as a whole, and then identifying a manageable sample from that large group. A single, well-designed interactive video programme can address the general needs of a wide audience, but there are always limitations. It's no good even trying to design a programme about two disparate subjects to answer the needs of three special interest groups. A programme that tries to address too diverse an audience is likely to fall between stools and succeed with no one.

If the interactive video programme fits into an established pattern (of teaching, perhaps, or advertising) it is likely that its audience has already been defined, whether formally for some previous exercise, or simply through experience. If interactive video is a new departure entirely, some serious market research could be the extra expense that pays for itself in the long run.

If the audience is already defined, well and good. It may then be that one or more members of the production team have done similar work in other media, and know their audience well. But remember that the trial-and-error method of audience analysis can be an expensive one, if it takes several programmes and as many searching postmortems to arrive at a working formula for subsequent projects. Here as elsewhere, skimping on preparations is false economy.

A survey, however informal, is one effective way of learning about your audience and their expectations. This could mean a round of visits to at least some of the people you want to reach, and/or their immediate superiors (students and/or their teachers, for example, or workers and/or their supervisors). Alternatively, it could justify a detailed survey designed and conducted entirely by professionals.

Many organisations feel that a survey is something they can conduct themselves, relatively cheaply and easily. This is usually only true if the survey is designed and administered by someone who is both qualified and experienced. A survey of any size and subtlety is extremely tricky to design and administer if it is to be representative, valid and ultimately illuminating. A survey badly conducted can stir up suspicion and resentment, waste time and money, and still produce information of no real practical value. A survey is an excellent way to draw up an audience profile, and to sample opinion – but it has to be done well. It is a field of consultancy which can be well worth the money.

The sorts of questions you need to ask in an audience analysis (obviously) vary according to the nature of the job you are doing. But some basic points are universally applicable. Consider, for instance:

* What sort of educational background does your audience have? Primary, secondary or tertiary? Academic, technical or commercial? Arts or science?

* What is likely to be their attitude toward formal instruction? How long is it since they left school? Does their work require them to study or attend training courses?

* Do they use instruction manuals regularly? What other ways do they acquire information they need in their work?

* Do they usually work under supervision? Who supervises them, and how?

* Do they generally work independently, in co-operative groups, in isolation, or in the company of other people doing other jobs?

* What are their leisure habits? Do they go out drinking in the evenings, play bingo, have hobbies? Do they watch television, go to films, read books? Which papers do they read? What kind of music do they like?

* What age group are you considering? Are you likely to be dealing with one sex more often than the other? Are you addressing any special cultural groups? Does your audience have a marked political bias? Where do they live? How much money do they earn?

* What prior knowledge do they have of the subject you are going to present?

* Why do they want to participate in this programme? (In fact, do they want to participate in this programme?)

* How familiar are they with new technology?

This kind of list, of course, could go on and on. Some of these questions may seem odd, but knowing the answers to them all could significantly affect not only the content, but also the style of your presentation. A programme aimed primarily at male accountants in their early forties, who live in detached houses in the suburbs and play golf on Sundays is not going to be couched in the same way as one aimed at young factory workers in the inner city, or one which addresses an audience who, regardless of background, speak English as a second language.

Your target audience will not always be so clearly defined as this; however, general trends are likely to emerge which will help to shape the presentation of the programme, and ease many creative decisions. The style of language, the choice of people who appear on the screen, the pace of the programme, the music employed–all these are things which affect how your audience receives the message. Subtle points of style profoundly influence the user's attitude toward the programme. People often remember an experience in terms of style rather than content, and this is something which the good designer learns to use to advantage.

THE MUSEUM PROJECT

Let us consider the likely audience of the hypothetical project for the State Museum of Fine Arts, introduced in Chapter 28.

The Museum undertook a formal survey as part of the application for the funding of this project. Questionnaires were handed out at public lectures and visitors were interviewed entering and leaving the building. A selected number of people using the research facilities, and a cross- section of Museum staff, were also interviewed about their needs and expectations.

This survey produced a analysis for use in this project and in many other aspects of the Museum's work. Visitors were found to fall into several distinct groups. Many people–school parties, foreign tourists, day trippers–are on the visit of a lifetime. Others–students, interested amateurs, workers from nearby offices–come regularly and participate in many Museum activities. Among other things, the survey found that some researchers who only occasionally work at the Museum are frustrated by the complexity of its printed catalogue, and that staff in most departments feel isolated from their colleagues and would like to know more about the work of the Museum as a whole.

Many decisions made during the design of the project will be based on this information. The kiosks in the main entrance hall, for instance, are likely to be used primarily by first-time and casual visitors, most of whom are tourists, few of whom have any formal acquaintance with art history, and many of whom are not fluent in English. This part of the programme will be presented in a simple and entertaining way, with lots of menus and options to allow users to choose information freely.

This programme will be offered in five languages (English, French, German, Italian and Japanese). Visitors will first see a brief guided tour of the Museum, conducted by an engaging television presenter, and will then be free to watch any number of short clips on art and artists. To wrap up, visitors can consult a gallery guide, a calendar of events, and a list of Museum publications if they wish. Since this last option is the one part of the programme that regular visitors are likely to use, the guide, the calendar and the publications list can each be called up directly without going through other segments.

And so on. While the visual information remains pretty much the same throughout, the choice of language will differ greatly in the staff training segments - especially those addressing curators with degrees in specialised fields. Different words and music will be used with the same video material to address adult and child visitors. People using the catalogue segment of the disc will have more detailed control options and devices than those looking at the public information programme.

This is not a comprehensive analysis of our imaginary project, but only an indication of the number and nature of the decisions which are made in such a programme, and the types of information which can help to make them.

CHAPTER 35: FLOWCHARTING

The difference between interactive video and the linear medium is characterised by the addition of one significant item to the documentation on which the programme is based. Traditionally, the seminal documents in any video project are the script and the storyboard. (These are discussed in more detail in Chapter 37; basically, the script lays down the words, and the storyboard, the pictures, more or less as they are to be in the finished programme.)

Interactive video, the fruit of video and computer convergence, adds to these two one important document from the world of computer programming – the flowchart. The flowchart is a sort of map, which represents in graphic form not so much where things are, but, rather, how things happen. (We still talk about 'place' in a flowchart or a programme, in rather the same way we talk about the arrangement of material on a videodisc in terms of 'real estate'. It is not the physical location of the recorded information on the tape or disc which is important, but the point at which it appears in the information programme.)

On paper, a flowchart is typically a network of geometric shapes and connecting lines arranged in some meaningful way. The simplest of flowcharts interprets a linear sequence – a conventional video programme, for instance – as a series of boxes of similar size and shape, each linked to its immediate neighbour by a short, straight line.

Each of these represents one discrete component of the whole – one scene in the video programme, perhaps. The boxes can be distinguished one from another in virtually any way, by numbers, keywords or any other reference which fits logically into the documentation of the project. This kind of flowchart is useful in planning of all kinds, for it shows structure at a glance, and explains in the simplest terms the relationship of one element to another.

As the flowchart grows more specific, symbols of different shapes and sizes can be used to distinguish components of different kinds from one another: screen-shaped boxes for moving footage, for instance, and crisp rectangles for still frames and artwork.

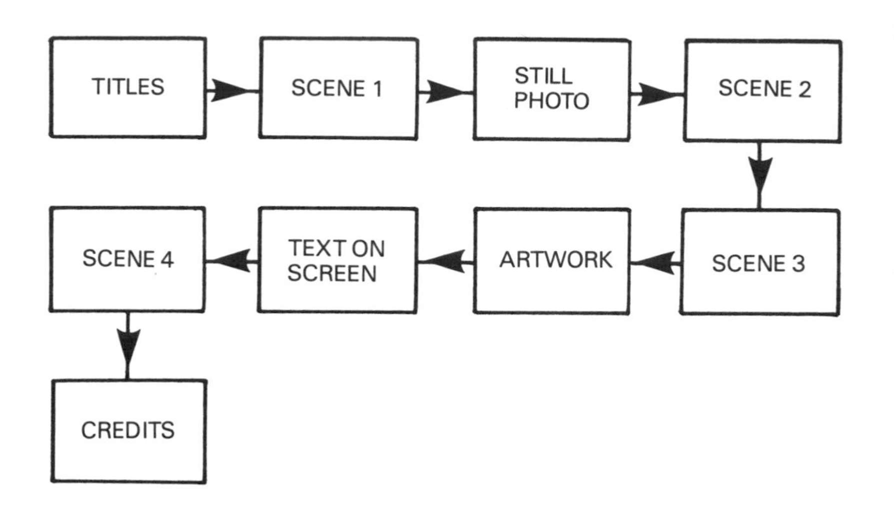

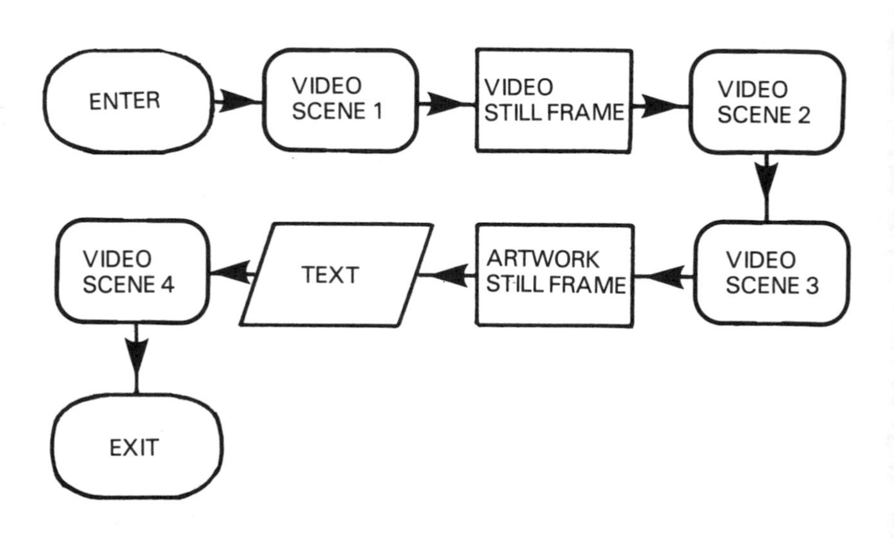

TWO SIMPLE FLOWCHARTS

This is how computer flowcharts are composed, with different symbols for different elements in the program. There are both international and national standards to dictate what any one symbol represents. In other words, there is no one universal standard. The same symbol does not always mean the same thing, and usages unique to one programmer or organisation may crop up from time to time. 'For this reason, great care (and often great resourcefulness) is needed in reading the flowcharts of anyone whose personal whims are not publicly accepted; considering that the flowchart is a major tool in the science of communication, it can sometimes be strangely uncommunicative.''[1]

The emergence of interactive video complicates this already volatile issue, for the addition of video sound and pictures to a computer program means the addition of a new set of symbols to the flowchart. So, just as standards of flowcharting vary, so there is no one way of representing an interactive video programme which will be immediately and universally understood even by the cognoscenti. If there ever were an opportunity to establish a standard graphic language for flowcharting interactive video, it was lost early on, as pioneers both commercial and academic began to work independently of one another.

Certain organisations with a need for standard notation have established their own systems, some of which are widely recognised. In straightforward computer programming, for example, companies such as IBM provide flowcharting standards (and aids, such as stencils and templates) which many smaller organisations have adopted gratefully. In interactive video, companies with plants capable of pressing Level 2 discs, and consultants to do programming for clients without their own resources, also offer both standard notation and aids such as guidelines and templates to help inexperienced authors prepare flowcharts.

The same principles apply to flowcharting computer programs and interactive video programmes, and the basic options are represented in much the same way by most widely-accepted systems of notation. However, it is critically important to know the precise meaning of every symbol and convention on any flowchart, or risk a serious misreading. In production terms, this means that everyone who works with the flowchart at any time must understand its conventions and vocabulary.

The flowchart for a fairly simple interactive video programme could be prepared by someone entirely without formal training or experience – as could any of the documents leading up to the execution of the programme. Many a pilot project is undertaken on a modest scale by a production unit well versed in conventional video but unused to the computing side of the project, working in a new medium without the benefit of experience or consultancy.

If an authoring system is being used, the whole programme may be produced in-house using existing human and material resources. Alternatively, the flowchart may be prepared by authors and/or designers without programming skills, and then handed over to a programmer for translation into computer language. In either case, the flowchart must be clear, complete and logical if the person who ultimately sets down instructions for the computer is to produce an effective program from it.

So while neither training nor experience is essential, at least a spot of consultation with someone experienced in the preparation of flowcharts is likely to save the first-time designer a great deal of time and effort. Flowcharting is a skill – anyone can learn to do it, some people will have a natural flair, and most people will get better, more confident and more creative with practice.

The flowchart for an interactive video programme should identify:

* Every segment of moving footage.

* Every still frame.

* Every screen of computer-generated text and/or graphics.

* All the branching options being offered the user.

* The choices the user might make at each menu, index or decision point, and the consequences of these.

* Every point at which the programme can be entered or left.

In computing, a distinction is made between a program flowchart, which shows how data flows through a specific computer program, and a systems flowchart, which shows the construction of an information handling system of which any one computer program may be a single element. The systems flowchart is like the plan which shows the location of different buildings within a compound, while the program flowchart is like a blueprint of any one of those buildings. The systems flowchart summarises the whole, while the program flowchart describes each part.

In our hypothetical project for the State Museum of Fine Arts, the flowchart which shows the relationship between the catalogue, the public information programme, and the various staff training programmes, illustrates the design of the system as a whole. Each of these independent aspects of the project represents a separate computer program module, each with its own program flowchart.

The flowchart on which the computer program is based usually evolves in two stages. First, an outline flowchart is roughed out, which translates the brief or the proposal on which the project is based into a series of discrete components arranged in some meaningful way. This ensures that all specifications are met, all needs satisfied and all components incorporated into the grand design before the work of detailing begins.

From the outline flowchart, it should be clear how the various parts of the project fit together, what relationship they bear one another, where any given path leads and when any segment is likely to appear. Mistakes, ambiguities and discrepancies are identified and corrected at this stage. Here, too, there are likely to be revisions or amendments to the original specification as the work of translating words and ideas into operations further clarifies the objectives and means of executing the project.

Once the outline flowchart is finished and agreed, work begins on a detailed flowchart, which takes the definition of each operation to the level that the programmer needs to complete the exact work of writing instructions for the computer. How detailed this actually is depends on the way in which the computer is being programmed, whether through an authoring system or in a computer language. The various authoring and computer languages work rather like different human languages, each with its own vocabulary and syntax—the programmer, like the polyglot, sometimes needs more information to make a statement clearly and unambiguously in one language than in another.

While there is no standard graphic language for the design of flowcharts, many basic symbols and concepts are widely applicable. By way of example, let us consider a few familiar symbols, and their possible meanings:

In a Level 2 programme, a piece of moving footage on the video screen.

In a Level 3 programme, material on the video screen, moving or still.

In Level 2, a still frame. In Level 3, computer-generated text or graphics.

The fixed or maximum time any item or series of items will be displayed.

A decision point

A test

Input from the user

An entry or exit point, where the user can begin, end or interrupt the programme.

A printout

For example, this might be a typical sequence in a Level 2 programme. (The explanations, of course, would not appear in a real flowchart.)

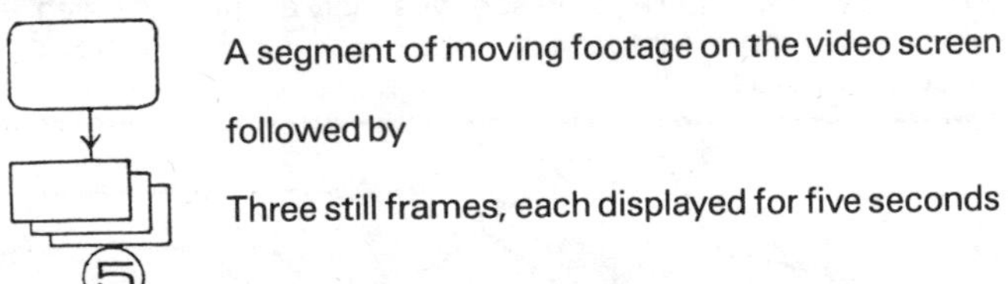

Every discrete component in the programme represents a separate 'information block', be it a single still or computer screen, or a segment of moving footage which runs for several minutes. Eventually, in the post-production stage, every frame of video material will be given a unique reference number, and every piece of computer-generated text and/or graphics assigned a unique 'address' on the floppy diskette. These are the numbers used to identify the start and end of each information block in the computer program.

In the interim, some less specific system of identification is needed to distinguish the separate information blocks during the pre-production and production stages. Naturally, there are various ways of going about this. A combination of letters and numbers is one good method. The letters logically are the initials of various keywords, such as:

M Moving footage on the video screen

S Still or freeze frame on the video screen

G Computer-generated graphic

T Computer-generated text

I Index or menu

Q Question

P Printout

Even after the information blocks have been arranged sequentially in the outline flowchart, it is good practice to number them in increments of at least ten to allow for extra material which may be incorporated subsequently. The first segments of moving footage, therefore, might be identified as M10, M20, M30 and so on, while computer screens could be G10, G20, G30 . . . or T10, T20, T30 . . . depending on whether they contained graphics or text.

This, of course, is only one approach—the authoring system, established practice or logic inherent in the programme may dictate some other way of distinguishing separate information blocks. The important thing is to apply one system clearly and consistently, whatever its principles.

By way of example, let us consider some basic patterns for testing. The first is a 'pre-test'—that is, a question is posed to determine whether the user need see the upcoming segment. This could be an actual test question which the user would have to answer correctly to avoid seeing the next lesson, or simply an inquiry which lets the user decide whether to pause for more information or to move on. On the flowchart, this sequence could be represented as:

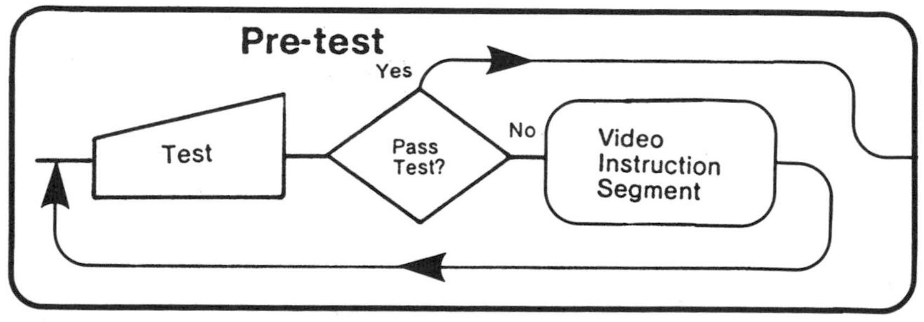

The second is a 'post-test', in which the user is quizzed at the end of a segment before being allowed to go further. The user who passes moves on directly. The one who does not is shown a remedial segment and tested again. In this example, the choice is simply between a right and a wrong answer, and the user is expected to proceed fairly quickly. In more complex situations, of course, the pattern of testing and branching would be much more complex.

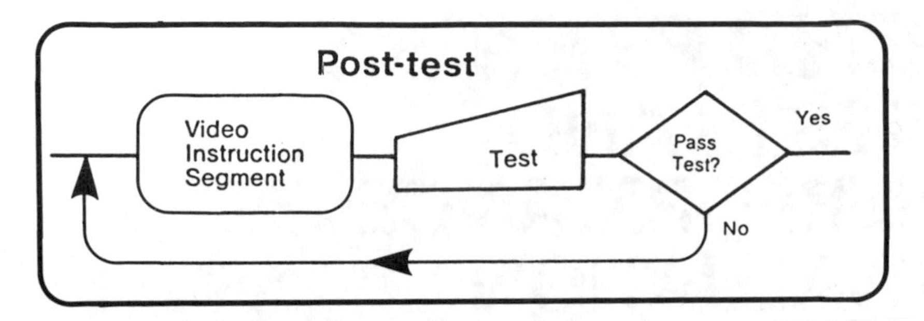

A outline flowchart may well fit onto a single page, while a detailed flowchart could paper a wall. The size and complexity of the flowchart reflects the size and complexity of the project; it also reflects how the minds of its designers work, just as the final computer program does. Flowcharts are put designed in different ways, too–you may go through piles of paper, use a chalkboard, cut out shapes in coloured paper, stick bits of felt to a board or doodle on the backs of envelopes.

What is certain is that many ideas and configurations will be tried and found wanting, and some of those which do survive will be modified beyond recognition by the time they are finally recorded in the completed flowchart. The preparation of the flowchart is one of those jobs which will probably take a long time; however, when it is done well, it represents the watershed in the production schedule. With a good flowchart behind you, you begin to move away from the gruelling design stage, and into the actual execution of the project, where decisions at last are easier to make.

1 Chandor, Anthony, et al. The Penguin Dictionary of Computers. Penguin, London, 1977. Page 179.

CHAPTER 36: TESTS AND PERFORMANCE EVALUATION

The idea of tests and evaluation immediately brings to mind teaching and training, but the principle can be applied to a wide range of applications—and not only to the people using the programme, but to the programme itself.

REVIEW AND REMEDIATION

Frequent short tests, however informal, are one structured way of offering the user a chance to review or seek assistance. If the user answers a specific question unsatisfactorily, or performs poorly overall, the programme can branch in any of a number of ways:

* The whole segment can be played again, just as it appeared the first time.

* A single, relevant part of the segment can be shown again.

* The original video footage can be presented with a different audio commentary, or extra help from text, graphics and captions.

* An entirely new segment can present the information again with different imagery, both visual and verbal.

It is possible and often advisable to offer more than one route (again, automatically through the computer program). Failing on the first attempt, the user can take a quick look back over the original material and then try the test again.

The one who passes moves on, while the one who clearly isn't succeeding can see a remedial sequence.

TESTING

As we said, testing evaluates not only the user, but also the programme - cumulative results can identify weak points in both. If it becomes apparent that everyone has problems with the same material, the fault is more likely to be with the programme than the users.

Testing is an integral part of the instructional scheme. With the flexibility that interactivity offers, decisions on what kinds of test to use, and when to give tests, can be as important as the test questions themselves. The different kinds of tests can be used variously to add variety to the lesson as well as to evaluate different kinds of learning.

Common types of test are:

* **Dichotomous** –True/False, Yes/No, Agree/Disagree–in which the user simply decides one way or another about a statement. A disadvantage with this type of question is that the user has a 50% chance of being correct. This type of question tends to depend on clear statements of fact, as well, which can be limiting.

* **Multiple Choice**,in which the student has to select the best answer from a choice of several. This type of question is more challenging and versatile than a straight true/false choice. The thing to remember when designing multiple choice questions is not to give the answer away by phrasing it in more positive terms, making it longer or shorter or in any other way conspicuous from the incorrect answers, or placing it in the same position or in some discernible pattern throughout the programme.

* **A scale**, which lies somewhere between the first two. Users are asked to select a number on a scale (say, 1 to 10), the two ends of which represent opposite extremes. This may require a subjective judgement which is extremely difficult to grade or evaluate.

* **Ranking**, in which the user must arrange all the points in list in some specified order. This is perhaps the most challenging of the lot, for the user must evaluate all the points in a list, and the relationship of each to the others. In this type of question it is particularly important to ensure that the points are mutually exclusive. It can be extremely frustrating to attempt to list, in order, points which overlap.

* **Matching**, in which the student must make up pairs from two random lists of words or statements. It is important here, too, that each point be clear, and unambiguous, or the student may be confused by conflicting choices.

* **Missing Words**, or the 'closed' form, in which the student has to 'fill in the blank' in a list or statement. Here it is important not to give the answer away in the phrasing of the statement, and to avoid such hints as providing the exact number of spaces for the right answer to be entered. The user can be given a choice of possible answers and asked to pick one, although this option, too, tends to make it 'easier'.

* **Keywords**, or the 'open' form, in which the user's answer must contain one or more of the 'key' words used in the lesson. This type of question is difficult to structure, and even more difficult to evaluate, for the designer has to anticipate every possible correct response.

Of course, tests are not always straightforward. Sometimes correct spelling is implicitly part of a correct answer, while sometimes it is more productive to make allowances for spelling or typing errors. Sometimes, too, there is more than one acceptable answer to a question. Most authoring systems allow for these kinds of considerations, but anticipating them is not easy.

Testing can come anywhere in a programme (not just at the end of a lesson) and it is worth thinking about the best place to put tests of different kinds.

* If the information is fairly basic, or may have been learnt before, it if often helpful to put a test before a lesson, so the user who is already familiar with the material can move past it more quickly. This is particularly useful in introductory courses, or those addressing a wide audience. There is little benefit in teaching people what they already know.

* In a long or complex lesson, periodic short tests both help to break up the flow, and give the user a chance to pause and review in a structured way before moving on. If the user does begin to falter, the programme can immediately branch to an appropriate remedial segment, instead of allowing the user to carry on well after the original thread has been lost.

PERFORMANCE EVALUATION

Tests can of course be used to the traditional end of evaluating the user's comprehension of the lesson. Many authoring systems record not only the user's answers to questions, but the time taken over each test, and the number of remedial sequences seen (in fact, a complete record of the user's progress through the programme), and can print out that information.

From this kind of record, an instructor can identify areas where a student would benefit from some alternative form of study. Interactive video isn't going to put good teachers and trainers out of work: on the contrary, it promises to help put their skills and experience to the best use, making programmes and giving tuition while the hardware handles the bulk of the basic, routine work.

Of course, the user may not be a student on a training programme, but someone unknown to the evaluator, who stopped at a point-of-sale unit or a consumer education display and played with a short programme there. In such a case, the user's performance gives the designers of the system a good idea of how it is working—whether as they intended, or for better or worse. Some systems monitor use overall without evaluating individual user's performance.

THE MUSEUM PROJECT

It is such a system that we will be using for our Museum project. The computers used in the public information programme have the auxiliary function of surveying the needs of one large group of the Museum's visitors. Of course not all visitors will use the system, but those who do, both by their expressed interest in certain collections, and by their answers to a few simple questions at the end of the presentation, will tell the Museum something about their backgrounds and interests. The question of evaluation as a whole is discussed in Chapter 33, together with that of setting objectives.

Meanwhile, the staff training programme will evaluate both the progress of individual trainees and the effectiveness of the this method of training. Real testing will be used only in this portion of the programme. Questions will be posed in the menus of the catalogue and the public information programme, but only to help guide users through the material available to them. The brief 'questionnaire' at the end of the public programme is meant to garner feedback from the users of the system, not to quiz visitors on their knowledge of art history.

All the types of questions described above will be used in the staff induction programme. These include a series of true/false statements about the set-up of the Museum itself ('All Museum employee are civil servants – true or false?'), an exercise in matching painters and movements ('Colville, Holman Hunt, Duchamp . . .', 'Dada, Photographic Realism, Pre-Raphaelite Brotherhood . . .'), a list of emergency procedures to be ranked in order ('Evacuate gallery', 'Sound alarm', 'Shut doors' . . .), a typical catalogue entry to be completed by filling in blanks, and so forth.

THE VALUE OF TESTING

This is a contentious issue in itself, which we will not dare attempt to address here. Of course, not all questions have hard and fast answers. Tests are not necessarily a true reflection of ability. The computer catchphrase, 'garbage in, garbage out' has parallels in instruction design: bad tests will produce bad results.

You may use tests simply for review and remediation, or even to provoke users into thinking in different ways about the material being presented. Because you have new technology, does not mean that you have to use it to the exclusion of conventional training aids such as workbooks, for instance, or group learning. There is certainly room for the old and the new, which people sometimes forget in their rush toward the newer and newer.

Ironically, two of the fields where interactivity is most effective are, on the one hand, technical skills training and the teaching of pure sciences, where answers are usually either right or wrong, and 'soft' skills (interpersonal skills or management skills), where there is rarely anything more precise than a 'best' answer. That paradox should tell us that tests and scores have their place, but that good instructional design can work on many levels.

CHAPTER 37: DOCUMENTATION

Documentation describes the whole, large, sometimes dull business of record-keeping. Anything recorded on paper, tape or disk which reflects or influences the development of the project, is a part of its documentation. In the development of an interactive video programme, this includes:

* Correspondence.
* Minutes of meetings.
* Internal memoranda.
* Contact reports.
* The brief.
* The proposal.
* The treatment.
* The budget and budget updates.
* The contract.
* Work planners.
* Scripts.
* Storyboards.
* Flowcharts.
* Computer programs.
* The production log.
* The shooting schedule.

Worse yet, this is only the tip of the iceberg. As well as reference and source material, there are endless lists of people, places and things, jobs to do and goals to meet.

It is important to keep track of any information that may be needed during the course of the project and, critically, of all the decisions made about it. This is far easier said than done, for accurate, concise record-keeping is a discipline only perfected with time. Documentation, like budgeting, is something many people – especially on the 'creative' side of the project – find dull and unrewarding. But the value of good record-keeping is undeniable. It is only too easy to base a decision on undocumented discussions which cannot be explained or defended later.

In interactive video, where the production schedule may be long, and where the programme may change and grow even after the production phase ends (with new computer programs, for exarnple), it is vital that the history of the project be adequately recorded. People may well come into the project once it is underway, who will rely entirely on the written and oral history of what has gone before. Oral history is valuable, but the written record is paramount.

Of course, a quick, confidential debriefing is often the only way of passing on information which, for a variety of reasons, cannot be committed to paper–stories of personality clashes, for instance, or narrowly-averted disasters. Such human dramas do affect the project, but the people involved are hardly likely to write notes to file about them, except in the most veiled language. We have to acknowledge, in a chapter about documentation, that some of the important information inevitably goes unrecorded.

At the same time, it is hardly practical for every member of the team to record the process through which fundamental creative ideas evolve. Documentation refers primarily to material decisions rather than creative discussions. It may be cynical to regard documentation entirely in the light of self-defence, but that is sometimes the motivation behind the note to file or the action point in the minutes.

Good record-keeping is something of a natural aptitude–which some people simply do not have. If everyone on the project is going to keep abreast of the paperwork, the emphasis (if not the rule) must favour the report confined to a single-sided sheet of paper. Bullet points and lists, if they are comprehensive and unambiguous, are quite as good as fluent prose, and often easier to assess at a glance.

Documentation falls into two basic categories:

* Project documentation.
* Programme documentation.

The first describes the paperwork involved in the design and execution of the project itself, from the contract documents to the invoices. The second describes a part of the 'deliverable'–that which is handed over to the client at the end of the project. This is, as it were, the 'owner's manual'. Let us now briefly consider some of these key documents in the growth of the interactive video project.

THE BRIEF, THE PROPOSAL AND THE TREATMENT

Terminology varies, but for our purposes, these words describe the first steps in the design of the programme. Each will likely be drafted many, many times before being finally agreed. The short answer here may be, invest in a word processor–it will prove a valuable member of the production team, without which the person who actually types the project documents may not stay the distance. Certainly the use of a word processor encourages tidy, consistent and legible documentation in a way that cut-and-paste techniques and handwritten annotations do not.

Essentially, the brief is the document prepared by the client (those people commissioning the programme) to tell to the production company (those making the programme) what they think they want. If the job is being done in-house, the brief goes directly to the production unit and may be prepared with their co-operation. If the job is put out to tender, companies in competition for the work should all prepare their pitches on the basis of the same brief.

The brief ought to state the objectives of the project in the precise, practical terms described in Chapter 33. It should also indicate for whom the programme is being made–the character of the organisation as a whole, and the nature of the intended audience (described as we suggested in Chapter 34). It should suggest at least a ballpark figure (or a ceiling) for the budget. If equipment and facilities are available which can (or, more pointedly, must) be used, these should be described. If the brief is a good one, a number of fundamental decisions will already have been reached by the time it is agreed.

The proposal is the production unit's reply to the brief. It sets out how the production team propose to make the programme, what it will look like, what it will contain, what resources are needed, roughly how much it will cost. In interactive video, it should include a rough flowchart.

These two documents are unlikely to be exchanged through the post – one or many meetings may be required before either the brief or the proposal is presented in final form. It may even be that discussions stop at this stage, and the client either goes to another production company, or shelves the project altogether, whether to rework it later or to abandon it altogether. The production company may write off the costs incurred up to this point or may charge the client for the work. This is reasonable, especially in a field like interactive video, but it is an agreement which must be reached between the client and the production company before work starts.

If the proposal is accepted and the client (or its equivalent) awards the job to the production company (or its equivalent), then work begins in earnest on the treatment. Work on the flowchart (described in Chapter 35) may begin here too, for the flowchart develops alongside the treatment, script and storyboard. The treatment is the document on which the detailed flowchart, the script and the storyboard are based. It brings the design of the programme to the point from which these working documents can be developed.

Another fairly important piece of paper needs to be agreed and signed at this time – the contract. It sets down precisely the terms under which the two parties are agreeing to work together. It addresses questions like payment schedules, confidentiality and so forth, and must be thought through, discussed and revised as carefully as any of the production documents.

It is likely that the project will change between the first draft of the brief and the final draft of the treatment. As we suggested in Chapter 33, clear thinking in concrete terms usually leads to revisions and refinements. However, as we also observed, by the time these documents are completed and agreed, a part of the job will be done. As each milestone is reached, the way becomes a little easier.

MEETINGS, MINUTES, MEMOS AND CONTACT REPORTS

Meetings may occur at any point, between any number of people. A great deal of time and temper can be wasted in meetings, especially when there are too many, wrongly timed, or between the wrong people. On the other hand, not having enough meetings between the right people can lead to serious breakdowns in communication, and all the problems that this entails. If everyone has worked together before, on a project of a similar nature, there are likely to be far fewer meetings than if everyone and everything is new and untested. Striking the right balance is one of the features of good project management, the criteria for which vary from project to project.

In conventional production terms, there is a point at which the client simply must trust the production company. But there must still be regular communication between the two. In practice, the right balance is usually a combination of:

* Telephone calls and correspondence.

* Informal meetings between people closely involved with the project.

* Larger, formal 'milestone' meetings, often involving people at senior level, at key points in the schedule.

The importance of the telephone should not be under-rated, nor that of the informal meeting, whether set up at short notice, or held over lunch or a drink. Critical decisions, information, advice or comments can come out of discussion at this level, which sometimes fail to be recorded because the meeting was informal. Good records include not only the minutes of formal meetings, but memoranda and 'contact reports' noting the salient points of casual ones.

Minutes and memos need not record every word or idea expressed, only those germane to the development of the project. They should include specific 'action points' where a decision has been made for someone to do something. They should also note where a decision affects some document such as the budget or script. The notes of a meeting may include the minority decision on a volatile issue which the majority eventually carried. Early in the production stage, meetings are often the fertile source of ideas which, while they may not immediately be incorporated into the project under discussion, will ultimately bear fruit elsewhere.

The memo or contact report may be a note to file, handwritten in point form, so long as it records information for which no other documentation exists – and so long as it reaches people whose work may be affected by the information it contains. It is an excellent idea, for many reasons, to keep a strict record of who said what to whom, when and where. The file comprising correspondence, agenda, minutes and contact reports represents a documentary history of the project, and can be as valuable as the script, the storyboard or the flowchart.

WORK PLANNERS

Any number of forms may be designed to organise information, time and resources. One valuable form is a work planner that interprets the production schedule in graphic form on a single sheet. A sample of one such form is provided here.

WORK PLANNER Project: Client:	Starting date:									Completion date:										
	MONTH 1				MONTH 2				MONTH 3				MONTH 4				MONTH 5			

The idea is to list all the jobs which have to be done, in roughly chronological order, and to block out the weeks in the production schedule assigned to each. Each column in the month should be assigned the date of the day (usually, the Monday) on which the working week begins. Some months, of course, will have five weeks rather than four. It may be useful to add a column to list the people involved in every stage of the programme, or to make other notes – this is only a model of the kind of standard form which can help impose discipline on a large or complex job.

The work planner, like the production schedule, will be revised more than once. Like the flowchart, it may exist in more and less detailed forms – an overall work planner for the project as a whole, and detailed ones for different stages in production. The salient point is that a simple analysis of work to be done, confined to a single sheet of paper, is often valuable out of all proportion to its size or complexity in helping to make work more manageable.

THE SCRIPT AND STORYBOARD

As we noted in the last chapter, the seminal documents in the production of most film and video programmes are the script and the storyboard. The script lays down the words, and the storyboard, the pictures, more or less or exactly as they appear in the finished programme.

The degree to which the script and the storyboard dictate the shape of the finished programme varies enormously from production to production. Sometimes, every shot is worked out in meticulous detail, and every word agreed, before production begins. The shoot and the edit are then largely a matter of executing instructions recorded on tablets of stone. Sometimes, though, the programme is built around material which is more or less spontaneous – unrehearsed interviews, for example, or live action. Then, the script and storyboard are usually a combination of instructions and guidelines, describing the shape of the programme but not every word and picture that goes into it. One script may provide every participant with word-perfect speeches, while another just lists 'Points to be made' as prompts for the interviewer or speaker.

In conventional production, the script is often just words, set out in two or more columns. These include at least:

* **Audio** – the script proper, words spoken either by people appearing on the screen or by an off-screen narrator; and

* **Video** – a brief description of the scene, location or background, and such stills or captions as may be used, as they appear.

The script may refer to the storyboard, or, in the case of interactive video, to the flowchart. It may include a column for timings, or notes, or any other information that may be relevant or useful. What auxiliary information the script contains, and the way in which it is presented, are largely a reflection of the working practice of the production unit.

The script for an interactive video programme may be more involved than that for a linear programme. People who work with scripts, storyboards and flowcharts can pick out the shape of the final programme from its preliminary documentation in much the same way that people accustomed to working with blueprints can construct a model in the mind's eye from a diagram. These people need special information to read the script of an interactive video programme.

The script for an interactive video programme need not be read in the traditional manner, one scene following another in immutable consecutive order. Some key to the flowchart or the branching pattern must be incorporated into the script to help people use it in a way which realistically represents the construction of the final programme.

A sample page of such a script, based on our hypothetical project for the State Museum of Fine Arts, is shown here. It includes a frame from the storyboard, and screens for teletext and computer-generated text and graphics, as well as text for both audio tracks. (The painters it discusses are, by the way, as fictitious as the Museum itself.)

TELETEXT	VIDEODISC	COMPUTER OVERLAY	VIDEO	AUDIO 1	AUDIO 2
(Both) Room 12			Presenter in Italian Renaissance Gallery	Here we find paintings by the masters of the Italian Renaissance	This fine collection concentrates on Northern Italian schools from 1445 to 1520..
(Public) Full colour poster on sale in Gallery Shop		(Staff) Chronology of life of Ambroglini	MS Presenter in front of Ambroglini's Four Seasons	This is one of our most famous works - The Four Seasons, by the Florentine painter Ambroglini . . .	Ambroglini's Four Seasons was seminal in the shift toward classical subjects . . .
			CU Four Seasons	The painting portrays the personifications of the four seasons . . .	Note the delicacy of the colours and the neo-classical values inherent in the composition . .
			BCU Spring	The figure of Spring, for example, wears spring wildflowers and carries a newborn lamb . .	The facial modelling is entirely characteristic of Ambroglini's style in this period . . .
		(Staff) Points of resemblance to work of Pinotti the Younger	CU X-ray of The Death of St Anne	When this painting was X-rayed, an earlier work on religious themes was found beneath . . .	X-rays reveal an earlier painting, possibly the lost Death of St Anne by Pinotti the Younger . .
			MLS Presenter in Italian Renaissance Gallery	Let us look briefly at some of the other work in this gallery . . .	The Museum has a rare collection of Sienese work from the last quarter of the 15th century . . .

The script usually exists in several forms—which is where the word processor pays for itself again. People involved only in specific parts of the programme do not need or want to see the whole script. Performers such as the presenter and/or narrator are usually only interested in the bits they have to deliver. In an interactive programme such as the one we are discussing hypothetically, there will be separate scripts for different parts of the programme. In our example, the public information programme and the staff training material are quite independent of one another, and will be scripted separately, by different people.

If a cueing device (that is, automated idiot card) is used, a special script will have to be prepared for it, by the production unit or the people providing the service. This usually means putting the scripted material, free of distractions such as stage directions, on to a long roll of paper which is then fed through an autocue machine.

The shooting script—the version from which original video (and audio) material is prepared—is often the penultimate draft. There may never be a final draft which accurately records every word spoken. If there is, it probably enjoys the benefit of hindsight, and editing that reflects both changes to the scripted commentary and the exact words used in unscripted sections, once these are recorded on the final edit. At the very least, you will go through a lot of paper before the script finally passes into the dead files.

THE PRODUCTION LOG

The production log itself may comprise quite a variety of documents and forms. It is in a sense a book of lists—the who, what, when, where, why and how of the project.

On the video side, the production log might list all the still frames used (or even considered), and record salient information about each—source, medium, content and so on—in some systematic way. It might list the facilities houses used, with information about each, and a record of what work was done where. Ultimately, the production log is a combination of data and diary, a record of the production underway and a source of information for later jobs. Like the file which holds both minutes and contact reports, it is likely to be a collection which ranges widely in content.

The 'shooting schedule' is just one example of ephemera which arise out of a production job. When the team actually go out to shoot video material, it is with a schedule which includes not only the order in which shots will be taken, and the time ascribed to each, but also details like the full names of everyone involved in the shoot (and, usually, their telephone numbers), maps and directions, information about transportation and overnight accommodation as appropriate, lists of props and equipment, and so on.

All this information is important, and all of it must be correct. Even something as relatively small in the grand scale as the shooting schedule takes time and research to compile. The video production side alone is replete with paperwork - with copyright and release forms, contracts, lists, quotes, bookings and heaven knows what. The documentation of the computer program, whether it is executed through an authoring system or written directly in a computer language, can be as voluminous.

PROGRAMME DOCUMENTATION

This is, as we said, the 'owner's manual' delivered to the client at the end of the production stage, when the programme goes into use. It includes, for example:

* The contents of the video disc or tape, listed frame-by-frame.

* A detailed flowchart with indexes pointing to the places in the computer program where individual parts of flowchart have been implemented.

The authoring system may have more features than are used on any one project: the documentation should indicate which have been employed. This record of how the programme was made is important to the people who may subsequently want to change or adapt it. Information may be updated or amended over time. Programmes addressing a wide audience, such as generic courseware, often include 'windows' into the computer program to allow individual organisations to 'customise' the software. If people other than the programme's original designers are to effect such changes, it is vital than they have adequate documentation on which to base their modifications.

The sheer volume of documentation on a project of any size or complexity is one argument for farming work out to a production company, especially if the structure does not exist internally to handle this kind of work. It is also the argument for careful attention to project management, which brings us to the next stage in our discussion: the project team itself.

CHAPTER 38:
THE PRODUCTION TEAM

How many people does it take to make video interactive? That line sounds like the lead-in to an ethnic joke. Some of the answers you'll hear to it are fairly hilarious, too.

The size of the production team, like that of the budget, ought to reflect the scale and complexity of the project. Not that it always will – getting the numbers right is a problem in jobs of all kinds, and neither the video nor the computer industry is a model of sedate project management. Anyone who works in either field knows of jobs that went wrong – of very small companies producing monumental work, of teams so large that talented people sat idle, of projects so mismanaged that the full weight of the project fell on one person (usually, your informant).

Teamwork is a combination of talent, hard work and good planning, and is a function of management and personality as well as experience and training. Interactive video is still on such a growth curve that there are not enough people with qualifications such as formal training and practical experience to meet the demand. This may be the case for a few years yet, which means both that there will be many opportunities for people wanting to break in to the business, and a danger of some clients' falling into the hands of production companies who can't deliver the goods.

From the client's point of view, one way to guard against that is to approach a production company in exactly the same manner as a facilities house, or any other organisation offering technical services: by asking to see samples of work done for previous clients – and ensuring that at least most of the same people who produced that work you like are still around. (It's no use being knocked out by a programme made by someone who has since gone to 'the competition'.)

From the production company's point of view, the problem is larger, especially since interactive video is sufficiently new, and different from conventional disciplines, to pose problems in its novelty alone. Those people with genuine interactive production experience have the pick of jobs (and salaries) and the rest have to fall back on a combination of relevant experience and a willingness to learn the trade.

Ultimately, aptitude is as important as experience. People who have worked in video, computing or training are likely to have skills to apply to interactive technology. In a project of any size, there are enough specialised jobs that each expert can stay near familiar territory, providing that each can depend on the complementary skills of the others. It is in small projects that people have to become all-rounders, and it can be difficult to find staff with the requisite combination of skills.

In the end, it is easier to define the jobs that need to be done than to specify who will do them. In the course of a long project, one person may wear several hats, and a good many people are likely to pick up new feathers for the caps they were wearing when they came in.

Let us consider the elements of a good production team by discussing areas of responsibility and job titles that might appear on a large project with an equally large budget and a long production schedule. (Bear in mind that, in the real world, only a few production teams have the resources we are suggesting here.)

THE PROBLEMS ANALYST

One of the key people in computing is the systems analyst. This is the person who analyses how something is done, or ought to be done, and then (ideally) designs and implements new and better ways of proceeding. This job, obviously, did not spring into being with the birth of the computer industry in the middle years of this century, but is about as old as human endeavour. The job of the systems analyst is based on computers, but is ultimately one of problem-solving. For this reason, in interactive technology we prefer to call this person a problems analyst.

The problems analyst is first and foremost a communicator, and needs good verbal and written communications skills. This implies an orderly mind and settled working habits. The problems analyst is the member of the team principally responsible for setting the aims and objectives described in Chapter 33, and for laying down the fundamental design of the programme as a whole.

This means establishing and meeting a functional specification, helping to implement the programme design, monitoring content and quality control. The problems analyst is the architect of the programme, and like the systems analyst, 'must ally a creative imagination with a willingness to undertake considerable drudgery at times.'[1]

We might describe the structure of an interactive video production team in a relatively non-hierarchical way by comparing it to a wheel, with the problems analyst at the hub, and other individuals or teams at the end of each spoke. (Alternatively, in some projects at least, we might characterise the problems analyst as a spider at the centre of a web.)

The four chief spokes of the wheel are:

* Instructional design.
* Information programming.
* Audio-visual production.
* Project management.

Let us consider each of these in turn.

INSTRUCTIONAL DESIGN AND INFORMATION PROGRAMMING

Just as we call the person at the hub of the wheel a problems analyst rather than a systems analyst, we call the combination of audio, video and computer-generated

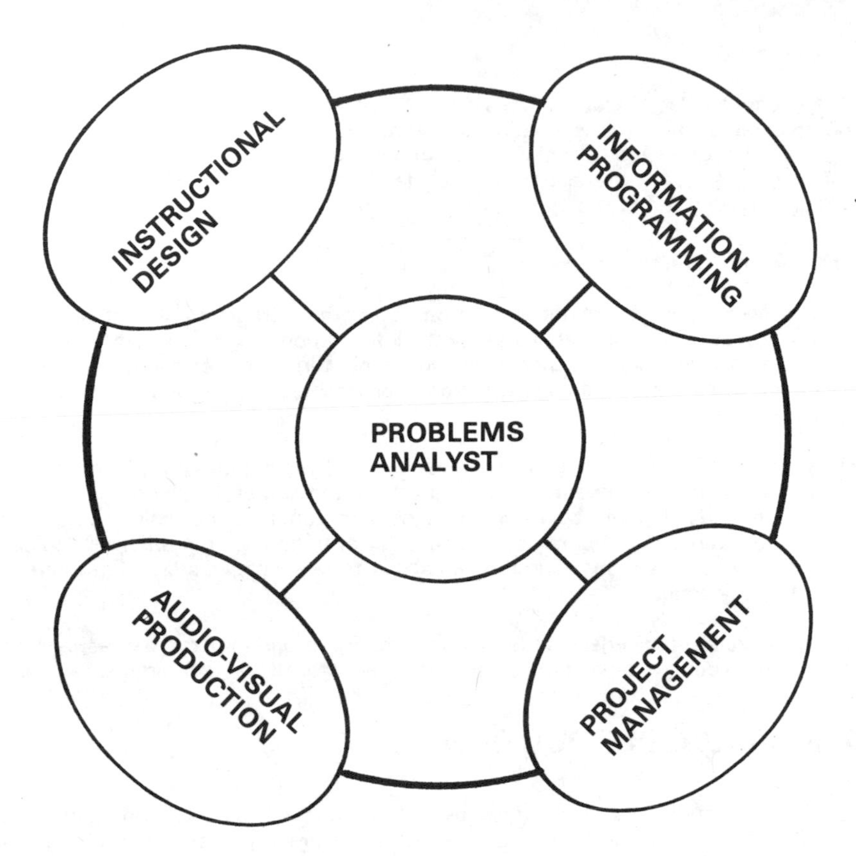

THE INTERACTIVE PRODUCTION TEAM

information and computer programs an 'information programme' to show that it is a much different beast to a conventional computer program or audio-visual production. The problems analyst is responsible for the design of the information programme as a whole, but the work of detailing and implementing this design falls between the instructional designer(s) and the information programmer(s) *per se*.

The Instructional Designer

All interactive technology is concerned with information transfer–in training, marketing or any other application. The instructional designer is someone who knows how to arrange and present that information in a way that its audience will enjoy and understand. This requires both an understanding of how people learn, and the ability to analyse information and its audience astutely.

The instructional designer works closely with the researcher and the scriptwriter, arranging the information to be presented into manageable segments. The instructional designer also works with the video director and art director, deciding how best to present each information block, and shaping the style and flow of the programme. The instructional designer arranges these information blocks into the pattern of the detailed flowchart.

While interactive video production does impose its own strictures, a sound background in the preparation of formal education or information transfer of some other kinds, touches on most of the requisite skills. Instructional design is a broad discipline, and the person who does the job will fall back on the specific expertise of other members of the team during the course of the project.

The Information Programmer

Information programmes are different from computer programs in concept and execution as well as spelling. Yet ultimately the information programmer translates the work of the instructional designer into computing terms, either for the industrial standard videodisc player's onboard microprocessor or, more likely, for an external computer.

The information programmer needs a solid background in computer programming, and will probably also have or develop a thorough understanding of at least one authoring system. (More and more, interactive video projects are turning to authoring systems in preference to computer programs.) A good information programmer is logical, methodical and conscientious – and has the ability to relate designs on a flowchart to events in the real world.

Together, the instructional designer(s) and information programmer(s) are responsible for the detail and construction of the computer-based information programme.

AUDIO-VISUAL PRODUCTION

AV production could, of course, mean as little or a much as the production of a sound/slide show or all the audio and video for an hour-long videodisc. However, as interactive technology most often employs traditional video techniques, it is on a typical video production team that we will concentrate here. Again, we are discussing the core of the team, and taking for granted that any number of specialists, such as technicians and engineers, will work on particular areas of production.

The Producer

In the absence of an executive producer, the producer has overall responsibility for both administration and editorial content. In the film world, the producer's role is usually one of financial administration, while in the television world, it is predominantly an editorial one. In interactive video, the producer has to take the rap for both.

It is conceivable that, within a large organisation, an executive producer will have responsibility for more than one project, each of which has a producer as well. In that case, all the various producers report to the executive producer, who has ultimate responsibility for everything.

The Director

In the world of broadcast television or corporate video, the director's responsibility is for the pictures that hit the screen. In interactive video production, while it may be the producer who agrees the overall objectives of the project with the client, it is the director who ensures that those intentions are fulfilled when the camera is rolling.

A director will normally have responsibility for both studio and location work, and the subsequent editing. But the director may well also be asked to take other responsibility—for content as well as execution. Sometimes one person acts as both director and the producer—often, someone who finds it impossible to delegate.

The Production Assistant, or PA

In television, the production assistant is the bright young thing who follows the director around with a clipboard and writes down everything from the shot list to the drinks order. In a small production unit, the PA might also be the project manager or, alternatively, the production secretary. Certainly, the PA is in the firing line, and virtually everything having to do with material questions—from insurance to continuity—comes to rest on the PA.

The PA usually keeps the shot list, but may also be responsible for anything from documentation of the project to keeping the camera crew supplied with change for the coffee machine. There are few jobs which vary more widely from project to project and company to company. The PA's job is sometimes set up as glamorous, but in the real world it is demanding, and requires skill and efficiency, an understanding of video, and supreme powers of diplomacy.

The Researcher and the Subject Expert

In a project of any size, there is a long trek between the acceptance of the initial proposal and the writing of the final script. What lies between, basically, is research.

It can take months for an intelligent person—who may well also be the scriptwriter—to talk to all the parties anxious to get their particular message across, to seek out the relevant information, and simply to get a feel for the project. A good researcher, although perhaps having no previous acquaintance with the subject, can find information that is needed. This is supremely a question of aptitude.

When the programme concentrates on a specialised subject, it is usual for at least one 'subject expert' (often from the client body) to act as consultant or adviser. This may be a professor emeritus or a machine operator, so long as it is someone who has a genuine understanding of the information to be transferred, who can both communicate that knowledge and comment constructively on the use made of it in the programme. The role of subject experts varies from project to project, depending on the degree to which the rest of the team is dependent on their expertise. The comments of subject experts are important not only in the design stage, but during the execution of the project as well—their advice may effect critical changes.

The Scriptwriter

Obviously, the scriptwriter's role is to come up with words that tell the story. However, the scriptwriter's job—in this team, particularly—is not the one it often is in broadcast and corporate video. Still less is it that popularly associated with the writer of screenplays.

Of course, interactive video can encompass virtually any genre from narrative documentary to real-time simulation, and a dramatisation, for instance, could well include scenes that would do credit to prime-time television. But even in as 'creative' an assignment as scripting a dramatisation, the writer is following guidelines laid down by others in the instructional design of the programme.

This position is not unique to interactive video production, for in projects of many types, the scriptwriter works closely with others who dictate the structure, if not the very words, that the programme will use. Advertising, training, and party political broadcasts are just a sample of the exercises in information transfer which involve the close co-operation of the scriptwriter, the director and/or producer, the art director and a subject expert of some kind. In interactive video production, this group is joined by the instructional designer, who plays a key role.

So the scriptwriter needs both to think and write creatively and still to be able to take direction, sometimes quite specifically, from others. Aside from a thorough understanding of conventional video production techniques, the scriptwriter must have the patience of Job and the vocabulary of Roget. Every word of the script is tied to the storyboard that describes the pictures and the flowchart that outlines the structure of the interactive programme. The three must always be co-ordinated, and a change to any one usually means a change to the other two as well. In many projects, the scriptwriter is responsible for teletext or computer-generated messages as well as the spoken word. With considerations like these, the process of honing and polishing, adapting and improving the script of an interactive video programme can be long and laborious.

The scriptwriter may also be the director, the researcher or the instructional designer and is certainly, as a wordsmith, dependent on the skills of all these people.

The Art Director and the Screen Designer

A project which has a lot of artwork—video frames and/or screens of computer-generated text and graphics—may employ both an art director and a screen designer. In a smaller project, one person may be responsible for both the design of the programme, and for the execution of individual screens.

The art director is involved from the first, working to set the right tone for the programme, and translating ideas into visual images. The art director draws up the storyboard that illustrates every scene of the programme, both moving footage and still frames. Whether the storyboard precedes the script or develops along with it, the two eventually tie together. The art director is responsible for this, and for the visual quality of the programme as a whole.

The artwork or design for individual screens may be executed by the art director, or by one or more screen designers and graphic artists. As we noted in Chapter 30, screen design is specialised work, and there are many gifted commercial artists who have no idea of how to work with a TV or computer screen. Interactive technology imposes its own rules, and the execution of computer-generated text and graphics differs from that of artwork in traditional media. Fortunately, here at least, people with a background in broadcast television or corporate video, or computer graphics design, have the requisite job skills.

PROJECT MANAGEMENT

This describes both the specific job of project management and the host of support services—finance, administration and so on—which must be supplied either directly within the production team or from the larger organisation of which it is a part. We will discuss two key figures here: the production secretary and the project manager. They may fulfil other administrative functions and, on a small project, may even be one and the same.

Just bear in mind that this is one area in which there may be people whose jobs are not discussed here, but who are an essential part of a project of virtually any kind. Here, they are accountants, office workers and the like; elsewhere, they will be camera crew, studio technicians and so forth. Our discussion concentrates simply on the core of the interactive production team.

The Production Secretary and the Project Manager

A key figure in the administration here as in many projects and offices is the production secretary. This job requires secretarial skills of a special kind—including, in this case, a basic understanding of interactive technology, a high degree of organisation, and nerves of steel. The production secretary usually does a variety of jobs, only some of which reflect traditional secretarial roles, and is under pressure most of the time. It's a good job for someone who likes hard work—and it is a traditional stepping stone toward other production jobs.

The average interactive video project involves such an intricate production schedule, so many people, and such a variety of facilities and services, that there has to be a single person responsible for all administrative and material considerations from paperclips to airline flights. This is the job of a manager (in television drama, that of the location manager or production manager.)

In an interactive video project, the job is called project manager. The project manager's job falls somewhere between the video producer and the production secretary, and might even be shared between those two. The project manager has to deal with both people and things, and needs energy, organisation, and a variety of aptitudes including money sense and communications skills. At whatever level it is filled, the job of project manager is a critical one. Without the project manager, who would sign the bar tab?

ESPRIT DE CORPS

Every member of the team brings to it specific creative, technical or managerial expertise. But everyone's job depends on the work of the others, even though it is often the case that few members of the team have a deep understanding of the disciplines or the limitations of the others' work. This is a fact of life for every member of the team, from the project manager to the screen desginer. It can be intensely frustrating to work with people who, even if they are sympathetic, do not fully understand what can and cannot be done within one's own field of expertise. Every member of the team will feel this—sometimes intensely—from time to time.

Vital qualifications for everyone—from the executive producer to the production secretary—are team spirit, and the ability to communicate effectively and work efficiently with other people. Heaven knows the production team is often rich in strong personalities; what is important in such a social activity as making an interactive video programme is to ensure that everyone on the project is a team player.

1 Chandor, Anthony, et al. The Penguin Dictionary of Computers. Penguin, London, 1977. Page 90.

CHAPTER 39: STAGES IN PRODUCTION

Two of the first questions people ask about interactive video are 'How long does it take?' and ''How much does it cost?'' To both questions, the answer is another question: 'How long is a piece of string?'

Costs and budgets will be addressed in the next section, largely theoretically, because it is impossible to put a price tag on anything which can vary so enormously in size and complexity as an interactive video production. This chapter addresses a somewhat more concrete subject – the actual work involved in making an interactive video programme – but it is still a discussion rather than a directive. Of course, certain points are universally applicable, but as with almost any human endeavour, different things will be done in different ways and at different times by different people.

A small in-house production unit, with their own facilities and a simple self-contained delivery system, making a short programme which is not substantially different from dozens they have made before, could turn out an interactive videotape inside a month. An equally experienced team, working on a highly technical project with a great many special features and problems, could spend two years making one programme. What it takes to make interactive video depends on many things, few of which have a fixed market value – the number of diverse elements in the programme, the skill and experience of the production team, the way in which the project is managed. Turn-around time at the disc pressing plant or video duplicating company is the least of considerations.

The stages in production vary and overlap in much the same way that the jobs on the production team do, depending on the size of the project and the resources at hand. The two groups that we identify as 'the client' and 'the production team' may be one and the same. The project may be produced entirely in-house, or may employ consultants, freelancers and independent production facilities. No one way of proceeding is inherently better than another – each project must be evaluated on its own grounds.

So, bearing in mind that any number of subjective factors will influence any one project, we can consider what follows simply as typical of interactive video production generally. Like most instructions, these points will probably not prepare you sufficiently for what happens when you actually make a programme yourself. This does not mean that making an interactive video programme is more difficult than, say, assembling a pre-fabricated garden shed, or learning French, or rearing children, but that instructions on paper tend to make things seem more orderly and straightforward than they prove in practice.

We are addressing here in a single chapter a subject which is matter enough for a book on its own. The point is not to teach you how to make interactive video on your own for fun and profit, but simply to outline the nature and dimension of the work involved in a typical project.

In fact, much of the planning and preparation that goes into an interactive video programme has already been described in Section VI. What follows on from that is, as we have said, largely a matter of executing work to a carefully detailed plan. This is one way of going about that work.

PRELIMINARY ANALYSIS

The very first stage in any project is often accomplished informally, or even subconsciously, as it becomes apparent that something must be done, some way of doing it must be agreed, and some action taken. The job could be anything from improving quality control in a factory to recording computer-generated data to selling products.

All interactive video programmes are ultimately solutions to problems – even discs from the home entertainment catalogue, which are addressing the problem of selling videodisc players to the domestic market. The first step in any project is recognising the problem – acknowledging that it exists, and then trying to analyse it. This may sound blindingly obvious, but it is often a very big step.

The problem has to be analysed even to be identified. This may be accomplished virtually without conscious effort, or may represent a job in itself, through anything from a brainstorming session to a survey. The work of setting the objectives and identifying the audience begins here – you need to know exactly what you want to do, and precisely whom you are addressing, to understand the problem fully. This is the work described in Chapters 33 and 34.

The conclusions reached at this stage may not be complete or even correct: detailed analysis in the later stages may bring a critical details to light. However, work to this level should be sufficient as a basis for the brief, or its equivalent – the preliminary document on which a proposal for the solution of the problem can be based. Both these documents, and the work they involve, are discussed in Chapter 37.

So, the stages leading up to the preparation of the brief include:

* Recognising that there is a problem.
* Identifying the problem.
* Analysing the problem in general terms.
* Setting the aims of the project.
* Identifying the likely audience.
* Determining what resources are at hand.
* Evaluating possible restrictions or qualifications.
* Estimating what money might be available.
* Drawing up the brief.

There has been no mention yet of interactive video – or even of video. There may be factors which dictate the selection of one medium over any others. However, if there are not, it would be a better exercise to take the preliminary analysis right up to the preparation of the brief entirely without bias toward the choice of medium, let alone toward any one delivery system.

There should be clear space for thinking between the brief and the proposal - even if some of the people preparing the brief think they have a pretty clear idea of what they want to to see in the proposal. One aim in preparing a proposal is to think creatively about possible solutions to the problem at hand. If the brief is set out without inherent bias toward one particular solution or approach, it may excite some ideas from the people preparing the proposal which are entirely new to those who prepared the brief. This is one strong argument both for going outside the organisation for creative suggestions, and for soliciting proposals from more than one creative consultant.

The main job of the people receiving the brief is to decide on a solution to the problem. The very first thing they ought to do – and many production companies fail to realise this in their eagerness to win commissions – is to look long and hard at the brief and pick it apart where it is flawed. If the brief has been prepared well, it is a sketch on which the design of the programme can be based. If it is sloppy or insubstantial, too ambitious or riddled with contradictions, it is no fit document on which to build. The consultancy that is really doing its job might just begin by tearing the brief apart and rebuilding it before trying to prepare a proposal. This is why the brief should not anticipate solutions – it may be over-reaching itself.

On the basis of a workable brief, the production company (or its equivalent) identify and evaluate possible solutions, proven and probable, to arrive at the one (or maybe two) that seem most likely to satisfy an optimum combination of the stated objectives and conditions. This means more research and analysis, both to arrive at a reasonable solution and to develop it sufficiently to test its worth.

In many interactive video projects, the work and the budgets involved are large enough to warrant a small-scale simulation of the final programme even at this stage. By the time the proposal is presented to the client, the first wave of production work should be complete. This is why so much as tendering for a large project can be an expensive business, and why production companies sometimes charge just to prepare a proposal, no matter what happens subsequently.

So, the jobs which face the group preparing the proposal include:

* Challenging or amending the brief where necessary.
* Identifying and evaluating types of approach - live presentation, print, film, video, tape/slide and so on.

For the sake of brevity, let us assume that the decision falls on interactive video. The work then becomes more specifically that of selecting a delivery system – the salient aspects of which are addressed in Section V – and choosing an approach to the material – a broad topic briefly addressed in Section VI. Bearing in mind that these bullet points encapsulate some fairly detailed paperwork, the final proposal should include:

* An outline of the proposed treatment.
* A design and description of the delivery system.
* Outline flowchart, script and storyboard.
* A preliminary budget estimate.
* A preliminary production schedule.
* A proposal for evaluation criteria.
* A description of the production team.
* Recommendations for an authoring system or computer program technique.

When the proposal is accepted, the more detailed work of preparing the treatment can begin. This, too, is described in Chapter 37. When the treatment is accepted, the actual work of producing the programme begins. Between these few short sentences, weeks and months of meetings, discussions, research and revisions may pass. The work involved here is that described in Section VI, briefly:

* Setting specific, practical objectives.
* Conducting a detailed audience analysis.
* Preparing the instructional design.
* Designing the video programme.
* Proposing a treatment for the project as a whole.

This is a point at which the stages in production begin to overlap. In a project of any complexity, even to draw up the treatment, it may be necessary to draw on extra resources. Certainly once the treatment is accepted, it will be time to:

* Draw up a production schedule.
* Prepare a budget.
* Assemble, recruit, seduce or shanghai people for the production team.
* Select and hire freelancers.

It is only here that the work traditionally associated with video production really begins. In practice, this may be weeks or months into the production schedule, depending on the complexity of the preliminary research and design work. The work is still confined to paper, and subject to any number of brainstormings, meetings, discussions and revisions. This is the design stage as it usually occurs in video production or computer programming. The team can now:

* Draft flowchart(s), script(s) and storyboard(s).
* Discuss, amend and agree flowchart(s), script(s) and storyboard(s).
* Prepare penultimate flowchart(s), script(s) and storyboard(s).
* Write accompanying print material (workbooks, manuals, and so on).
* Arrange for translation as necessary.

Once the indispensable triptych – the flowchart, the script and the storyboard - has been accepted, production work on the audio-visual side shifts gear dramatically, and begins to commit energy, time and money to executing ideas which have been confined to paper hitherto.

The production may involve original video footage with a presenter and/or narrator, interviews, dramatisations and so forth. The same preparations that go into the production of broadcast television must be made here, although production standards may be set higher. It is important to establish technical specifications early on, to avoid wasting time and money on material which is not up to scratch. If there are no in-house production facilities, or if these are not equal to the demands of the project, it will be necessary to hire technical facilities and expertise. Some existing visual and audio material may have come to light during earlier research; it is now time to assemble and prepare re-usable film, video, artwork and recordings.

These are the sorts of jobs traditionally associated with the pre-production stage in the video industry:

* Agree technical specifications for master videotape.
* Select and book facilities: video studio(s), sound studio(s), editing suite(s) and so on.
* Select and hire professional performers.
* Identify and brief other participants (interviewees, subject experts, guests and so on).
* Select and hire crew and technicians.
* Assemble costumes, props, sets and the like.
* Identify any re-usable visual or audio material available.
* Research archival video and audio material.
* Select music and/or sound effects.
* Clear copyright on existing material.

This last point, concerning copyright, is a reminder of the many small jobs that can only too easily be overlooked. There are lots more in both the video and computing sides of the project. An experienced video producer or computer programmer could double the length of this list – but instilling fear and despair in the heart of the novice is not the point of this book. Simply bear in mind that there are many little housekeeping jobs implicit in these general points – the sort of record-keeping discussed in Chapter 37, for instance, lies behind every stage in production. When you actually make an interactive video programme, many of these details will become apparent and many more, if you are employing an experienced production team, will be seen to by the professionals to whom they are familiar chores.

Housekeeping and documentation certainly become important as the audio-video side of the production moves into the production stage *per se*. This is popularly imagined to be the glamorous part of the project, although there are many people working in broadcast and corporate video production who would dispute that. It is now time to:

* Shoot original film or video footage.
* Prepare original artwork.
* Photograph original still frames.
* Prepare special effects.
* Transfer material from other source media onto videotape.
* Record voice-overs (including alternative language tracks), music and sound effects.

While all this is going on, the considerably quieter work of the instructional designer and information programmer is carrying on apace. In appearance, it is signally less dramatic – a matter now of honing and refining the directives laid down by the flowchart, script and storyboard into a series of simple, unambiguous instructions for the computer. Of course, if computer-generated text and/or graphics, or teletext, are involved, this work must be undertaken by now, too. The authoring system may be new or unfamiliar, as the hardware itself may be. Getting acquainted with the delivery system is also part of the job.

The members of the team preparing the information programme must now:

* Learn the authoring system.
* Master the delivery system hardware.
* Select and/or design and integrate purpose-built peripherals and control devices.

* Finalise instructional design.
* Design computer information program.

* Design computer-generated text and/or graphics.
* Prepare teletext.

Meanwhile, the video side has moved into post-production, and is assembling the material for the edit on videotape which is the master for interactive programmes on disc or tape. Editing is often accomplished in two stages: a preliminary (offline) edit, which is followed by extensive analysis and criticism before the final (online) edit is undertaken.

One important job – that of planning the arrangement of material recorded on the master tape – clearly involves the work of the instructional designer as much as that of the video director or producer. As we've said, efficient real estate management is the key to rapid random access and all that follows from that, so deciding how best to arrange video and audio material on the master tape is a job which brings the computing and video sides of the project together. This may happen earlier, but must happen before editing begins.

Once the master tape is complete, the vital frame reference numbers can be assigned to the tape and entered into the flowchart and computer program. The script and the storyboard can be revised one last time, to include details that may only have been finalised in the edit. Barring some last bug in the computer program, these three documents should now accurately reflect on paper the construction of the finished programme.

In interactive video, typical post-production jobs include:

* Plan layout of video, audio and control tracks on disc or tape.

* Assemble, review and approve off-line edit.
* Complete on-line edit.
* Assign frame reference numbers to master videotape.

* Enter frame reference numbers in flowchart and computer program.

* Input instructions for computer screens.
* Assemble information programme.
* Test and refine information programme.

* Complete final drafts of flowchart(s), script(s) and storyboards(s) and related documentation.

Here or earlier, it may be a good idea to simulate the running of the finished programme, wholly or in part, using available material and equipment. This may mean simply stepping through still frames on a slide projector, or could involve a relatively realistic simulation of some small part of the programme. Seeing some approximation of the work in progress helps everyone – production team and clients – visualise the finished work more clearly.

It is worth pointing out that the choice of delivery medium does considerably affect the production task. Videodisc offers such good reproduction, even several generations removed from source material, that it may dictate specific quality control criteria in selecting existing visual material and producing new footage and stills.

Also, the question of 'field dominance' is a technical nicety which greatly affects the finished product. Fields and frames were explained in Chapter 10 – basically, one frame of video comprises two separate, interlaced fields. If still and freeze frames are a part of the programme – which is likely, since they are one of disc's strengths – then special attention must be paid to 'correlating' the fields so that all material is recorded in the same way. Failure to do this will result in 'interfield flicker' or 'judder' when a freeze frame appears in the finished disc. (Philips were the first to use the 'field correlator' developed by Quantel to facilitate this.) This question should be discussed with the video editor and technicians.

Finally, presuming all technical specifications have been met and all instructional design and aesthetic criteria satisfied, the master videotape is ready for the last stage of the production. What happens here depends on the medium selected: tapes can be turned out relatively quickly at any reputable replication facility, while discs have to be mastered and pressed at one of a handful of plants so far concentrated in Japan, the US and north-western Europe. However, if the mastertape has been properly prepared, the discs can be turned out quickly and cheaply, even if they may have to travel some distance back from the pressing-plant, so the inconvenience is negligible.

If the delivery system is Level 3, the computer storage medium will have to be replicated also (this is likely to be a floppy diskette or perhaps an audio tape). There may be workbooks or instruction manuals as well, which is a print job in itself, and may be best off in the hands of someone with print buying experience. Another of the small but critical housekeeping jobs can be mentioned here, too: many people forget that the finished discs, tapes or diskettes will need dust-jackets and protective cases, which might as well have your label on them rather than those of the replicating company. If a workstation or other display unit is a part of the ultimate delivery system, it too must be designed and a prototype, at least, constructed.

So, we are considering jobs like:

* Send master tape to videotape replicating company or disc-pressing facility.
* Send master diskette(s) for replication.
* Review and approve check disc or tape, if possible.

* Arrange for labels, sleeves and/or jackets for delivery system software (tapes, discs, diskettes).

* Design and construct delivery system workstation or presentation unit(s).
* Produce accompanying print material.

* Deliver and/or distribute hardware and/or software.

Even after the completed tapes, disc and diskettes, the delivery system hardware and the supporting documentation are in the hands of proud new owners, work remains. You may want to send out a press release or go on the conference and exhibition circuit to show off your handiwork. You may want a stiff drink. It would be an excellent idea to pay your suppliers and invoice your client (if not necessarily in that order).

The very thought of starting in on the mountain of quotes, orders, invoices, delivery notes, expense claims, credit notes, petty cash slips and related financial paraphernalia that even a relatively small project accrues, is too much to contemplate here. Costs and budgets are discussed in the next section.

CHAPTER 40: YES, BUT HOW MUCH WILL IT COST?

How long is a piece of string?

There is no clear guide to what an interactive video programme costs to make. It's quite true and not very elucidating to say that an interactive video programme can cost a little or a lot, depending on the nature of the project. Barring the cost of buying the equipment and bringing production staff on stream, routine programmes on an in-house delivery system might be made for marginally more than ordinary videos. More likely, though, the cost of the project will reflect both production expenses which do not differ greatly from those of conventional linear video, and costs run up in a design stage which is considerably longer and more complex than any in the traditional medium.

The time taken to select and develop delivery and authoring systems, to research a project and to analyse its component parts, to prepare draft flowcharts, scripts, storyboards, computer programs and documentation, can account for more than half of the production schedule before a foot of video is shot. This phase is labour intensive, and potentially very expensive. The key is to regard interactive video production as a project rather than a programme.

Making interactive video, on disc or tape, may involve jobs and equipment unfamiliar to conventional video production, but these have market values, and the project is costed like any other – by estimating what costs will be incurred, getting competitive quotes for bought-in services, estimating in-house expenses, and calculating accordingly.

A few years ago, when every interactive video project was a pilot involving hours and hours of research and development work, there was scepticism about the value of the idea in the rough and tumble of the everyday market. But technology develops quickly, and changes within a single decade can make all the difference: interactive videodisc, once regarded as expensive and exotic, has been found to be an economical and cost-effective medium for many applications.

The drop in price of computing power over the 'seventies is a striking example of technology decreasing in price and improving in quality. It also provides one clue to the growth of interactive video as a commercial proposition. Even through years of high inflation, the cost of computing declined steadily over the last decade. At the same time, machines became smaller, more powerful and more versatile: many micros now offer more than the old mainframes did – at a fraction of the price. We now seem to be on the upward swing, as lower hardware costs have to be set against higher software development costs. But some new markets emerged during those years, one of which was interactive video.

The technology on the interactive video market has become dramatically cheaper and more accessible in this decade:

* While the price of computing technology fell in both absolute and relative terms, videotape equipment became relatively cheaper, and considerably better in quality. When the videodisc was introduced, it took the same path down to competitive commercial markets.

* Computer/video interface packages compatible with most popular equipment can now be bought as a package on the consumer market.

* 'Authoring systems'—software which allows people to write computer programs entirely in everyday language—mean that someone quite without computer programming experience can prepare an interactive video programme.

* It is possible to put together an interactive video system from video and computer equipment already at hand, using an interface package and an authoring system, or to buy a whole delivery system—player, computer, monitor and software—right off the shelf.

The cost-effectiveness of interactive video is evaluated more in the terms of computer-based training than those of conventional video. The working life of an interactive video programme is different from that of its traditional counterpart: its linear playing time (that is, the time it take to play from end to end without interruption) is no reflection of the time it may actually take to run in use.

Also, as with virtually any kind of project, there are more and less expensive ways of going about things. For example, a collection of still frames transferred from 35 mm slides is much cheaper and easier to produce than a computer graphics sequence of comparable accuracy and detail—and likely to be more informative as well. On a conceptual scale, the efficient arrangement of material within the programme can considerably improve access times, branching patterns, and the quantity of material that can be stored.

Undoubtedly, the advice of a qualified consultant can be the wisest investment in the budget. Good interactive video isn't easy to produce: no one said it was. But, well made, it can be tremendously versatile and undeniably cost- effective: it has proven its ability to sell, to persuade and to inform, and it is beginning to influence work patterns in fields from engineering to education.

There are certain to be people who will use the technology prodigally, and who won't see a return on their investment. However, interactive video so far seems to have steered pretty well clear of folly, perhaps because the expense encourages people to plan and work more carefully than they tend to do in straight video production.

In any new market some degree of shakedown is inevitable as the real leaders emerge from the many who jump on the bandwagon. The fierce competition between the manufacturers of rival hardware systems will be with us for some time: but this has not only the effect of complicating decisions about delivery systems, but also of keeping a sharp edge on prices and enhancements. Competition for business among production companies and consultants is rather less straightforward. Choosing a consultant is rather like selecting a facilities house or production service (topics addressed in Section VII, and again, from an economic perspective, in the next chapter.)

The only way to predict how much a specific project might set you back, is to do a rough estimate. This can be done, crudely, as soon as you have a broad idea of the work ahead–the size of the production team, the type of delivery system, the number and types of information blocks in the programme and so on - and can be refined as the brief becomes more detailed. The sorts of considerations which go into costing and budgeting are discussed in the next chapter. An example of a rough preliminary budget follows in the final chapter of this section.

CHAPTER 41:
PREPARING A BUDGET

The same basic considerations go into the budget of any project, whether you are planning a home improvement or an interactive videodisc. There are jobs to be done, and materials which are needed. Both goods and labour have a market value: by getting competitive quotes from the suppliers of the work and materials that you need, you can estimate the cost of the project that you wish to undertake, either as a whole or in its component parts.

If, for example, you want to convert the attic of your house into a study, you calculate the cost of the project first by dividing the work into stages (such as design, conversion and decoration) and the expenses into categories (such as labour, materials and tools), and then by getting estimates for work and goods that you are going to buy or hire. If you could do some of the work yourself, you might set the value of your time and effort against the cost of hiring someone to do the job for you.

Had you the time and the skills, you might do the whole job yourself, buying the materials at retail prices; very likely, in that case, you would already have most of the neccesary tools. On the other hand, you might want to design the conversion yourself, and to decorate the finished shell, but to hire labourers for the carpentry, wiring, plastering and so on. They would have their own tools, and could get materials at trade prices, but you would have to buy decorating supplies. If you were handing the whole thing over to professionals, the project would involve an architect and decorators as well as labourers.

An interactive video project is analysed in exactly the same way as that domestic conversion: by dividing the work into stages, and estimating the labour and production costs (in materials and facilities) likely to be incurred at each. (Typical stages in production are described in Section VII.) It may be possible to complete the whole project in-house, but quite likely some expertise and facilites will have to be hired, and certainly materials will have to be bought.

Cost analysis is usually conducted in three parts:

* The preliminary analysis is calculated broadly as a basis for contract negotiations, competitive bids and funding applications.

* The detailed budget which follows is the document which directs spending during the execution of the project, and against which its ultimate efficiency and profitability is evaluated. One of the first jobs once the contract is signed, is to rework the proposal, the budget and the schedule to reflect changes agreed between the client and the production company.

* The final appraisal determines how much the project really cost. This calculation is based on the budget, comparing actual costs to estimates, but may include things for which no allowance was made. Some unforeseeable expenses are inevitable, but the budget aims to obviate surprises in the final analysis.

Costs can be broken down in different ways, both to facilitate calculations, and to analyse where the greatest expenses are incurred.

* Usually, there is a division between internal and external costs ('above and below the line')—that is, between the people and facilities that can be found in-house, and those which have to be bought in.

* It is useful to distinguish between the design and production phases of the project. The first is time-consuming, labour-intensive, and difficult to cost accurately: people are thinking, planning, talking, researching, going back and forth between the drawing board and the discussion table. The second is fairly routine, and should be easier to cost: the work designed in the first phase is being executed according to explicit schedules and instructions.

The single most important element in preparing realistic schedules and budgets is a detailed breakdown of the costs you expect to incur. This may seem a statement of the obvious, but many, many projects come to grief over time and expenses for which no allowance was made in the planning stage.

MAKING A DETAILED BREAKDOWN

It is more difficult than it seems to estimate the expenses a project is likely to incur. This means breaking down the production work into every job that must be done, and the cost of materials and services into every discrete expense. This is best done in consultation with the people involved in each job, to get a clear idea of what work and expenses are involved. A practical budget is based on experience, quotes and advice; if you aren't confident that you have the first, you can at least scout around for the other two.

A breakdown works two ways, both analysing the work involved in any one job, and determining how much is required of any person, facility or supply. From the breakdown, it is possible to determine which people, facilities, equipment and resources are needed for an activity such as a day's shoot in a hired studio, or the transfer of film to videotape. Conversely, it is also possible to draw up a schedule for, say, a freelance video director, by noting when and where there is a call for that member of the team.

This breakdown should be made as early as possible in the project. It should be based on, or developed alongside, the production schedule described in Section VI, which lists the jobs to be done, in the order in which you plan to do them. Since the cost of labour is considerable, it is important to draw up a firm schedule at the same time as the budget—and to keep to both: one depends on the other. (The team tend to think of certain stages in the production as jobs that they know well and do not need to see minutely described, but the budget must list every single element which will incur an expense. Analysed from this perspective, the simplest project can look most formidable on paper.)

Collecting rates cards from production companies and facilities houses will give you both a good idea of what current market prices are, and of the kinds of jobs, goods and services for which charges are made. At some commercial production facilities, you will be charged for everything you do and use, down to the last fribble. It is important to anticipate charges of this kind in asking for quotes and in booking time. If your job is twice the size you said it was going to be, the facilities house will have no compunction in charging you at least twice as much as they first quoted.

You can find production companies and facilities listed in publications like Britain's Creative Handbook or in the advertisements and supplements in video trade magazines. A good collection of rates cards is a handy reference tool both to finding facilities and to costing jobs.

Both the cost and the labour involved in an interactive video project could be broken into three broad categories:

* Define the aims and objectives of the project and produce the overall design.

* Design and produce the audio-visual element on disc, tape or slides.

* Design the computer screens and implement the information programme(s).

The first phase is one of research and content analysis, the work described in Section VI. The cost of this stage is the most difficult to estimate, precisely because it is the first, and because it is labour-intensive. The work is that of study, consultation and design, with allowance for many revisions along the way. Even a seasoned production team can badly under-estimate the scope of work involved in a new project; here as much as anywhere, time and effort must be appraised and budgeted realistically.

A typical breakdown of this stage in cost terms might be:

PEOPLE

* Project management.
* Content or subject expert(s).
* Instructional designer(s).
* Authoring system expert(s).
* Screen and graphic designer(s).
* Members of the video production team.

* In a typical production unit, some of these people will be on staff, some will be hired specifically for the project, and very likely one or more subject experts will be seconded from the client or its equivalent.

The next phase of the project is the production of the audio-visual material. The cost analysis on which this part of the budget is based should list every single activity in the production schedule, citing the individuals whose work is needed, the time allotted to each task, and the facilities, equipment and supplies which will be needed. This could be prepared as a chart or a list, under headings which are further sub-divided into different categories. A typical analysis might include:

PEOPLE

* Members of the production team.
* Consultants, advisors and researchers.
* Presenters and artistes (including voices, performers and extras).
* Crew on location and in the studio.
* Editors and editing suite staff.
* Artists preparing graphics and artwork.
* Special services from translators, model makers, set builders, prop makers and so on.

* Whether provided internally or externally.
* Whether salaried, contracted or hired pro rata (by the hour, the day or the job).

FACILITIES

* Video studios for live action, rostrum camera work, captions and graphics.
* Sound studios for recording and transferring voice-overs, music and sound effects.
* Sites for location work.
* Facilities for the preparation and transfer of archival material, filmstrip and the like (telecine transfer, multiplexing).
* Editing suites for preparation of original video footage and assembly of master tape.

* Whether available in-house or hired (either as a package or pro rata).
* Basic facilities hire, plus cost of services.

EQUIPMENT

* Camera(s), lights, sound equipment and so forth, for a live shoot on video or film, in a studio or on location.
* Camera and lights for rostrum camera work.
* Equipment needed in the video and sound editing suites.

* Whether owned, purchased or hired (either as a package or pro rata).

STOCK

* 35 mm film and slide film.
* Video stock (1″ or U-matic) for shoots, transfer and editing.
* Processing.

GOODS

* Props and costumes, whether purchased, hired, borrowed or made, in-house or outside.
* Materials for set construction.
* Bits and pieces for the PA's bag (make-up and toiletries, first aid kit, emergency supplies).

ROYALTIES
* Hire or purchase of archival material.
* Music royalties.

INCIDENTALS
* Travel and transport (including vehicle hire and mileage allowances, where agreed).
* Meals, accommodation and expenses.

* Whether incurred attending meetings, on reconnaisance, shoots, or in production.

OVERHEADS
* Administrative and other in-house staff providing support services.
* Stationery, typewriters, word processors, photocopying.
* Telephones, telex and telegrams.
* Messengers, postage, courier services.
* Insurance, social security and health insurance.
* Carnet documents, visas, duty, taxes, bonds.
* Rent, rates, taxes, heat, electricity and so forth for offices.
* Overheads for in-house production unit.

TRANSFER
* Post-production and transfer to videodisc or videotape.
* Replication of copies.

Now we come to the last part of the project, information programming, where it all comes together. This is where the fine detailing in the design of both the individual computer screens and the branching pattern as a whole finally take place. After this, instructions to the computer are input, probably through an authoring system. This is then linked to the video player, and the ideas which began germinating so long ago in the early design phase are at last tested.

This phase also is also labour-intensive, but a little easier to budget than the first part of the project, since the team now have a reasonable idea of the general size and scope of the work involved. Typically, it might involve:

PEOPLE
* Project management.
* Instructional designer(s).
* Content expert(s).
* Scriptwriter(s).
* Information programmer(s).

EQUIPMENT
* Delivery system including:
  Computer.
  Video player and monitor.
  Authoring system.
  Furniture or work station.

* Office space for one or more complete work stations will be needed, and some equipment necessary to the developmental work may have to be leased.

This list is quite long enough, and yet it is still skimpy. All these things contribute to the cost of the project, and neglecting to appraise them in detail can soon wreak havoc on your budget and your schedule. In the next chapter, we will touch on some of the things which make for good budgets and efficient costing.

CHAPTER 42: KEY POINTS IN BUDGETING

Many people are loathe to discuss money: in the 'creative' environment, there is often a good deal of impatience with project managers and cost accountants who insist on agreeing the budget, and seeing quotes and receipts for every expense. But meticulous record-keeping, while a demanding and singularly uninteresting part of the job for some, is essential to the ultimate success of the project.

Preparation of the budget at the outset of the project, and the costing at its conclusion, can be surprisingly time-consuming activities. What is important is not so much financial acumen – there are usually people with those qualifications outside the project team – but discipline, to prepare a realistic budget, to stick to it, and to produce a comprehensive costing at the end of the project.

This is a function of skill, experience, attitude and motivation – but there are key points which apply to almost all projects, some of which are discussed below.

FINANCIAL MANAGEMENT

Responsibility for the budget usually rests with either the producer or the project manager, depending on how the project is structured. It will likely be calculated in consultation with the project team, and even administrators and financial controllers from other parts of the organisation. It is usually modelled on a combination of past experience, fresh quotes and common practice.

Individual members of the team (such as the art director, the video or film director or the PA) are responsible for keeping on schedule and within budget in the areas in which they have some autonomy in buying and hiring goods and services. Nevertheless, ultimate control should rest with a senior member of the project team, whose responsibility it is to keep a close eye on the budget, and to act when the project starts to go over budget.

It is also that person's responsibility to decide what to do when the project, or a part of it, does go over budget. One of the implicit requirements of the project manager and/or producer is practical experience in cutting corners discreetly, and bringing wayward projects into line.

It is important that every member of the project team who is spending money keeps accurate records. Get firm quotes in writing for everything you buy or hire. Distinguish between a 'pencil booking', which can be cancelled at no charge, and a 'firm booking', which carries a 'cancellation charge'. Get detailed receipts for everything.

It can be a good idea to ask the project team to complete weekly timesheets. However roughly, these indicate how any one person's time is spent at various stages in the project. They can be compared to the production schedule and the budget to indicate where work is going more or less smoothly than anticipated, and where the project diverges from the plan. They are important, too, if labour is going to be charged back to a client.

It is important both to keep within budget, and to document expenditure properly. Whether the project is a low-budget in-house training programme or a flashy point-of-sale unit, the project team must ultimately account for every penny spent. The invoice can be a challenging document in itself, and mistakes and oversights there—often occasioned by lack of comprehensive documentation—can cost you dearly.

IF YOU START TO GO OVER BUDGET

A detailed budget makes it easier to keep an eye on the mounting cost of the project, and to detect when and where work begins to go over budget. Short of coming clean and asking for more money, this is where the expertise of the production team comes to the fore.

Some of the options are:

* To substitute cheaper effects for dear ones (a still frame sequence for complex animations, say, or multi-plexing for videotape editing).
* To move work to cheaper facilities, or from an outside supplier or service onto in-house resources.
* To reallocate money from one part of the budget to another (perhaps necessitating, but allowing time to plan, economies further down the line).
* To drop or reduce some of the information blocks.
* To eliminate some expensive decorative element (such as the use of a well-known voice on the soundtrack, or the transfer of some awkward piece of archival footage).

Practical economies can usually be found, especially when the team as a whole goes looking for them. Here as elsewhere, it is important to make sure that everyone in the team knows when a change is made, and adapts accordingly.

HIRING VIDEO SERVICES AND FACILITIES

We noted earlier that an interactive video project incurs most of the same expenses as a linear programme on the video production side. As you saw in the list in the last chapter, most of the identifiable costs in the production phase are related to video: in any budget, the computer program is represented by hours and hours of labour, and the video programme by pages and pages of production costs.

As we have said, the ancient art of haggling isn't quite dead: never accept a company's rate card as being in any way indicative of what you will finally pay. If you are not careful, you could end up paying much more than the rates card would lead you to believe possible, but with bargaining and planning, you should pay less.

Video production may look like a money-spinner, but it is a fiercely competitive business, with more companies angling for work than there are clients to provide it. After all, you do implicitly pay facilities houses to make investments and take risks you don't want to touch, buying expensive equipment and kitting out fancy studios that may be outdated within months or years.

These companies want your business: let them bid for it. Every place you ask should be able to offer you something – reduced rates, extra goodies at no extra charge, a package deal on work carried out under one roof. You often have a choice between a large, comprehensive facilities house, and a number of smaller, specialised ones. The difference is based the quality of work, conditions, staff and rates at each, and biased by the large company's ability to offer attractive package deals. Of course, it can simply be easier to do most of the work at one place.

Once the negotiations are over, get quotes for everything, in writing. (It is important to confirm agreements in writing at every stage, but never more than where money is involved.) Be sure, too, that the quotes are explicit, and that you understand exactly what is included (and, thus, what you will still have to provide) each time. In a project with a long production schedule, it is important to ensure that prices on which you budgeted in January don't double by December: for the guarantee of the business, most suppliers are willing to make a deal on this.

LABOUR COSTS AND CHARGE-OUT RATES

The cost of using in-house staff can be calculated on the basis of 'charge-out' rates assigned to every member of the production team. Realistic charge-out rates are vital to a production company bidding for work on the open market, since they represent the figure charged per person per day to a client. However, they can also help an in-house unit evaluate project costs effectively.

Charge-out rates can be calculated variously, but usually represent an employee's annual salary (including holiday and sick pay), divided by the days in a working year (that is, 365 days less statutory and company holidays, and weekends – say, 250 days). This yields a basic rate per working day, to which some allowance for overheads must be added.

Allowance for profit can be made at this point, or as a percentage of the whole project cost. (A mark-up of 20% to 30% is about average in the video production industry – whatever they might tell you.)

There are various complex formulae for working out charge-out rates, overheads and profit margins. Most yield the same result as the rule of thumb which advises a ratio of 3:1 between charge-out rates and salary – the equivalent of the employee's pro rata salary should be charged once to cover pay, holidays, and so on, again to cover overheads, and a third time for profit. (Like most rules of thumb, this one is not scientific, but it's uncannily accurate.)

Rates for people whose services are being bought in are usually negotiated:

* Freelancers such as directors, scriptwriters or artists either accept a flat fee for the whole job (and, like you, hope the project keeps on schedule), or agree on a rate per project day, similar to those charged for employees.

* Presenters, performers and other artistes, however humble, usually have an agent, who will negotiate a fee and usually act as go-between, scheduling bookings and passing messages.

* Camera crews are often hired per diem (with allowance for mileage, overtime and other predictable expenses), at a rate which usually includes the hire of their equipment. The company that hires out crews has its own charge-out rates to cover the crew's salary or wages, the value of the equipment, overheads and profit. In a facilities house, it is common to offer a package deal that includes crews, equipment and facilities at one price.

We will say again and again, that all quotes are negotiable: minimum union rates exist for some jobs, and there is always a reasonable 'going rate', but video production is a highly competitive business, and when you have work to offer, you should be able to get it done at a good rate.

In Section VI, we discussed time schedules. Once a charge-out rate has been fixed for in-house staff, and rates agreed with freelancers and other bought-in personnel, the cost of labour can be calculated. Here, as everywhere, it is important to be realistic about the time it takes to do a job.

OVERTIME

Even if you are using in-house facilities, it is only professional to plan your time there as carefully as you would a booking for which you were paying by the hour—especially if other people are using the same facilities. It is only too easy to become cavalier about the real cost of using in-house facilities, and salaried and contracted staff. Running up a lot of overtime is expensive, whether you are paying double-time in a commercial studio, or working all weekend in your own production unit.

It is of course a great temptation, especially to companies competing for work, to budget on the best of all possible schedules. But failure to allow sufficient time in the design or production stages can lead to long hours, shoddy work, bad decisions and, inevitably, overtime. On the other hand, to err generously often means coming in ahead of schedule and/or under budget, and it could yield time and money for enhancements to the original brief.

CASH FLOW AND CONTINGENCY PLANNING

Cash flow can be a serious problem in an interactive video project: the production schedule can be a long one, and expenses accrue steadily. In both commercial and academic applications, most of the budget must be spent well before full payment or funding is received.

Depending who is funding the project, and how they are paying for it, reimbursement may come long, long after expenses have been incurred. Staff have to be paid, and overheads met; freelancers and small companies (most of the video production world) have acute cash flow problems of their own. A small company could go out of business even with a project worth half-a-million, if it is obliged to pay suppliers within thirty or sixty days, but cannot get payment from the client in less than six months.

It is, therefore, essential to ensure that money goes from the client to the production company as steadily as it is disbursed. A schedule of payments that covers your expenses is vital, no matter what the scale of the project or its application.

Certainly there are phases in the work which can be invoiced separately. For one thing, the industry has reached the stage where projects are so big, costly and complex that production companies can expect to be paid for drawing up detailed treatments, no matter what happens subsequently.

It is important that every change in the project be agreed between the production unit and the commissioning body before any work is undertaken, but where these changes may effect the budget, it is vital. In the heat of the moment, the budget is not necessarily the first thing one considers, but it is essential to keep it up to date with other aspects of the project.

Also, and again because the production schedule can be a long one, it is prudent to allow for contingencies such as changes to the brief, sharp price increases and unforseeable setbacks. The video production world is fast and furious: even reputable facilities can go under, or be taken over, on short notice–and sometimes at great expense and inconvenience to their clients.

These chapters have given you some idea how to go about costing a project–but every programme is different, and there is no set formula. Furthermore, most organisations have their own internal costing procedures which must be respected. The short answer is, ascribe a value to everything and, if anything, err generously. But let's now return to our hypothetical project, for a sample analysis.

CHAPTER 43: A SAMPLE COST ANALYSIS

Having just said that every project should be analysed down to the last paperclip, we now present a sample which is very rough indeed. We hasten to point out that this is only a broad outline, given in round figures approximated from London market rates in the autumn of 1983, and nothing like a reliable quote for an interactive video project.

It is dangerous to cite costs at all, for no matter how many warnings are posted around a ballpark figure, there are still people who will seize on it as definitive, and calculate accordingly. Therefore, this example comes flanked by qualifications: as we have seen, the cost of making video interactive depends principally on the scale of the production, and that can vary enormously.

This example is based on the hypothetical project that we have been discussing since Section V: the interactive videodisc commissioned by the imaginary State Museum of Fine Arts. As we have seen, it involves still and moving footage (from PAL and NTSC 1″ video, 35 mm cine film and 35 mm slides), location and studio work, archival material, still frame storage, a variety of background music, and dubbing from English into four other languages.

There will be a professional presenter, several voice-overs, interviews and vox pops with staff and visitors. With different programmes and soundtracks, the disc will be used to introduce the museum to visitors, to train curatorial and commercial staff, and to catalogue the museum's collection.

The production company will provide the producer, project manager, instructional designer, information programme designer, information programmer, production assistant and production secretary from its own staff. The director, scriptwriter and screen designer are freelancers hired by the production company for this project. Other people (crew, engineers, editors, artists, the presenter) will be hired pro rata. Consultants are drawn from the museum's staff, voice-overs will be spoken by a museum lecturer.

Exclusive of delivery system hardware and the authoring system, the project will cost about £150,000. Let us imagine that the producer and project manager prepare the initial cost analysis, taking advice (from the team and from the client) on the work involved and the time and facilities needed to complete each phase of the project. Bearing in mind that this budget is entirely hypothetical (and bracing ourselves for the dazzling shorthand of the facilities house invoice) the first draft might look something like this:

PRODUCTION TEAM

Charge-out rates (per diem) for:

Producer	£ 400
Project Manager	400
Instructional Designer	250
Information Programmer	250
Production Assistant (PA)	200
Production Secretary	100

Contracts agreed with:

Director	250
Scriptwriter	250
Screen Designer	250

1. PROJECT TEAM FOR THE DESIGN PHASE

Producer	1 month	£ 8,000	
Project Manager	2 weeks	4,000	
PA	1 month	4,000	
Production Secretary	2 weeks	1,000	
Director	10 days	2,500	
Scriptwriter	10 days	2,500	
			£22,000

3. LOCATION SHOOT ON 35 MM FILM (2 DAYS)

Producer	1 month	£ 8,000	
Project Manager	2 weeks	4,000	
PA	1 month	4,000	
Production Secretary	2 weeks	1,000	
Instructional Designer	1 month	5,000	
Information Programme Designer	2 weeks	2,500	
Screen Designer	1 month	5,000	
			£29,500

2. PROJECT TEAM FOR THE PRODUCTION PHASE

Crew (four people)	£ 1,000	
Hire of equipment	1,000	
35 mm film stock	800	
Processing and rush printing	4,000	
Presenter	600	
Editing (inc track laying, A&B roll production)	1,000	
Editor's expenses, black and white spacing, leader and dispensibles	100	
Neg cutting	800	
First graded show print	800	
Second graded show print	650	
Telecine transfer onto 1" PAL videotape	200	
Mix (10 hours)	800	
		£ 11,750

4. STUDIO SHOOT ON 1″ PAL VIDEOTAPE (1 DAY)

Studio hire	£ 500	
Studio crew (five people)	500	
Hire of equipment	300	
Set design	500	
Set construction	2,000	
1″ videotape stock	500	
Off-line videotape edit (2 days)	300	
On-line videotape edit (1 day)	2,400	
M&E sound mix	500	
		£ 7,500

5. PRODUCTION OF ART STILLS (2 WEEKS) £ 10,000

Involving:

Specialist photographer
35 mm colour slide stock, processing and mounting
Transfer to 35 mm filmstrip
Transfer to 1″ PAL videotape (plus stock 1″)

6. AUDIO-VISUAL SEQUENCE £ 4,000

Involving:

Photographer
Stock and processing
Original artwork
35 mm slide production
Slide programming and M&E sound mix
Multi-plexing to 1″ PAL videotape

7. ARCHIVAL FILM £ 4,000

Hire of archival footage
Telecine transfer onto 1″ PAL videotape

8. 1″ NTSC VIDEOTAPE SEQUENCE £ 500

Involving:

Hire and import of NTSC broadcast TV sequence from American network
Conversion to PAL

9. ARCHIVAL SOUND £ 500

Transfer of wax cylinder material

£ 4,500

10. TRANSLATION

Involving:

Translation of visitors' guide into French, German, Italian and Japanese
Hire of four readers
Recording of voice-overs in sound studio

11. MUSIC ROYALTIES

Royalties for pop music	£ 600
Royalties from standard music library	50

12. DELIVERY SYSTEM HARDWARE
(see also Section V)

Industrial standard PAL disc players	£1,200 each
Multi-standard colour monitors	400 each
Micro computers	2,000 each

13. DEVELOPING OR ACQUIRING AN AUTHORING SYSTEM

To build your own authoring system: Programmer (6–12 months)	£ 6,000–12,000
To lease an existing authoring system: Licence fees (per system)	£ 50–1,000

14. SCRIPTING DETAILED INFORMATION PROGRAMME

Instructional designer	3 months	£ 15,000

15. IMPLEMENTATION OF INFORMATION PROGRAMME

Information Programmer	3 months	£ 12,000

16. TESTING AND DELIVERY

Producer	1 week	£ 2,000	
Information Programmer	2 weeks	2,500	
			£ 4,500

17. PRODUCTION AND REPLICATION

5 master discs (English, French, German, Italian, Japanese)	£ 7,500
5 check discs	1,500

Checking and verification (5 days) by Production Team Programme Designer	1,250
25 copies (13 of English-language disc and 3 of each of the others)	500
	£9,350

This comes to just about £140,000. Depending on the terms of the contract, a 20% mark-up may be added to this, and the production team's expenses paid as well. In Britain, 15% VAT (Value Added Tax) will certainly be added. All that would nudge the price of the project to nearly £190,000—you see why this Museum was an imaginary one. The real point is that costs can quickly mount up, sometimes shockingly if, as a production unit, you are not properly prepared or, as a client, you have not been kept closely advised. Take note and, for the last time, be warned.

CHAPTER 44: SPECULATING ON THE FUTURE

"Vasi has no patience for people who learn to do something and then do it. He learns something and then he does something else."

"That's what science is," the Professor says.

– Max Apple, The Yogurt of Vasirin Kefirovsky

Looking at modern life, for good or ill, from the perspective of the writers of past generations, ours does seem an age of fantasy – perhaps not what Wells, Huxley, Orwell or even Beerbohm foresaw, but certainly a world very much changed from the one they knew. With our strange costumes, shiny machines and odd behaviour, we are 'the future' – at least as certain of our predecessors saw it.

People have always enjoyed imagining how life might be for other times and cultures. Just as Europeans once speculated about exotic peoples in foreign climes with faces in their chests and giant feet growing out of their heads like sunshades, so we have imagined 'little green men' and, latterly, 'extra-terrestrials'. We fantasize about 2001 and beyond just as our forebears fantasized about us.

Yet the salient point in the comparison of most speculative writing with subsequent reality, is that so much that really happens is never predicted. People tend to work forward by slow steps from the familiar to the likely, but change often comes (sometimes after long and uneventful gestation) from the arrival of the unforeseen, or the convergence – as in interactive video – of the apparently disparate.

Anyone who claims to know for certain what will happen to the world we know as time dissolves toward the year 2000 should be confined to the astrology pages of the Daily Planet. What we propose to do here is simply to follow a line of thought from what exists into what could be. This is by no means a treatise, but only a few open-ended ideas thrown in for leavening at the end of a long, factual tome.

Much of the fiction of the last century speculated about the likelihood of things which are taken for granted today: the great mystery was whether the imaginable were even possible. Now, the problem is not that we dream of that which is beyond us, but that technology is so far in advance of the social, commercial and political structures which might employ it, that much of what is possible is not altogether probable. There is often a hiatus between development and implementation, rarely for technical reasons. It is the material and social expense of replacing the old with the new that makes governments wary and investors cautious.

The great gap between the possible and the probable, between the products which are now selling well and those which are likely to affect our future, is hardly more apparent than in the field of interactive video technology. There are market forecasts aplenty that look forward five or ten years, but few commentators prepared to talk about the next century (indeed, the next millenium).

Already it is clear that non-broadcast video, cable and satellite transmissions are gaining on traditional broadcast television as a focus of our attention and our spending, and that computers are becoming a domestic commonplace. One American study predicts that the market for home entertainment equipment and services will double between 1982 and 1992, with the home computer market enjoying a walloping ninefold increase, and the videodisc market shooting up from US$0.2 billion to 1.5. Revenues from videotapes, cable TV and videogames are all expected at least to treble—and those from broadcast television to fall slightly.[1]

Another American study, on the interactive videodisc software market, noted that in 1982 spending was dominated by industry (59%), and by government and the military (29%), with education and information storage (7%) and point-of- sale (5%) running well behind. The same report predicted that the market which handled US$56m in 1982 would be worth twelve and a half times that much within five years, with point-of-sale units accounting for 30% of use outside the field of entertainment.[2]

The scepticism sometimes expressed on the consumer market is short-sighted: videodisc, and other elements of interactive technology, are certainly the force to be reckoned with in the coming years. The endorsement given the technology by the world's two largest computer companies, IBM and Digital Equipment Company (DEC), is a clear sign in itself.

Starting cautiously then, backed by facts and forecasts, it is certainly safe to say that established applications will flourish.

* The disc-based electronic manual will be common. Industries where training is of critical importance, and where there is money to spend, may lead the field, but the benefits of electronic manuals to all kinds of construction, manufacturing and maintenance jobs is undeniable, and their use is certain to spread.

* Training programmes based on interactive video will be used in all spheres of business and industry. The technology will contribute to teaching not only in schools and colleges, but also in the home.

* Marketing, both at point-of-sale and in its more subtle manifestations, will use interactive video routinely—and consumers will be accustomed to using interactive systems to look for bargains and ask for advice.

* Audio-visual presentations will move away from hand-operated overhead projectors and slide carousels into interactive technology and computer graphics: both the quality and the quantity of the visual information we use will change considerably in the next few years.

* And, of course, videodisc-based computer games will be popular in homes and places of public entertainment, and will become ever more intricate and elaborate. Games of mystery and logic will engage our interest as well as those requiring quick reflexes.

The key to it all is the integration of new technology into established patterns of information exchange. If we project beyond the electronic manuals, the point-of-sale units and the videogames of the immediate future, we can see ways and places where patterns will change in response to new tools and new attitudes toward both technology and information.

The idea of the 'home work station' has been bandied about for years, and grows more sophisticated with every new move on the market. It is certainly the end toward which these last few, safe predictions point. It is also already a commercial reality.

Real estate agents and commercial travellers whose home computers connect to a central terminal are implementing the idea in its simplest form. But the work station of the future will communicate not only with head office, but with any number of terminals, public and private, offering information and services of all kinds. It will receive signals, both analogue and digital, from broadcast radio and television, from cable TV, telephone lines and satellites. It will include not only the familiar VDU and a keyboard, but the hardware to receive, record, edit and display pictorial and textual material, audio signals and computer codes, to process electronic mail and to play games.

Data recorded digitally and transmitted by satellite or cable can travel great distances in time and space without loss of quality, to be reconstructed as pictures, text, graphics, sound, or even instructions to equipment. On a composite work station, information of virtually any kind, from any source, could be received, recorded, processed and displayed entirely within the station, transferred between media and re-broadcast – in its original or an amended form. A single processor could serve any number of terminals – 'a screen in every room' could displace a plethora of other hardware as the ultimate status symbol.

With complementary terminals in the places from which we seek information, goods and services, it would possible to do a great deal of the day's work and play from a single stationary position. We are on the verge of the great science fiction fantasy which sees the evolutionary end of homo sapiens as a sort of intelligent blanc mange propped in front of a screen.

While we are still ambulatory, let us consider just a few of the possible extensions of new communications technology into the not-too-distant future.

The steady progress through the various means of distance communication, from the inception of the penny post onward, has had a marked effect on the logistics of work, in elements as disparate as business travel, the use of stationery and interior design. But technology has so far not disbanded the business office. However, the cost of running an office (and of working in one) is high in itself, and many people could work as well, and live differently, if they did not have to travel to work.

There are obviously jobs in which people need to be together, but many people, even in large organisations, effectively work alone or in small groups as it is. For many, the office of the future could be a few centrally- appointed rooms, housing the main data terminal and a meeting place, and visited only when business demands.

It is difficult to imagine that shopping as a pastime will ever lose favour, for it is a social as well as a commercial activity. But day-to-day shopping is drudgery, and one of the jobs which still sharply divides the social roles of men and women. The logical extension of interactive video as a selling tool in stores and malls is interactive television as a medium for direct and catalogue sales—selling everything from groceries to hauté couture.

The consumer of the future, however isolated or housebound, could buy domestic staple items and luxury goods from all over the world without leaving home - browse through catalogues, compare prices, study product descriptions, watch demonstrations, order and even eventually pay entirely through a terminal. Staple goods could be ordered automatically, gourmets and people on special diets would have easy access to the food they need, Harrods' catalogue could sweep the world. Orders relayed directly from the home through the shop to wholesalers promise improved ordering, invoicing and quality assurance for consumer, retailer and wholesaler. The role of the retailer and the nature of the shop will change, too—and the role of the shopping centre as a social centre may have to be acknowledged more overtly than it sometimes is now.

Publishing is bound to change with the growth of optical disc technology—not only in media, but also in approach. There is a great deal to be said for the printed word—the book, as we suggested in Chapter 1, is a model of interactive information handling—but printing is an expensive and speculative business. Publishing on disc, and printing hard copies only on request—'Books While-U-Wait'—is an attractive prospect in many ways, even if it does cast a threatening shadow over the secondhand market.

The line between print and other media is already a fine one, for many bookshops and libraries now handle tapes and discs, and even computer software. The bookshop of the future may comprise a handful of bestsellers and antiquarian volumes set off by banks and banks of discs, a screen, a terminal and a printer - or even the hardware alone, with the databank at some distant central location. The asocial bibliophile could even stay at home, communicating with the libraries and publishers of the world through the work station.

It is conceivable that in the teaching of the future people will be able to 'buy' an Oxbridge or Ivy League education—lectures, lab work and all—as a set of discs for the home work station. Alternatively, one could compile an education with broadcasts and courseware from many sources. Even now, we have teletext and computer programs in the broadcast television signal, and enterprising stations are using off-air time to send programming to home video recorders and home computers. Distance learning needs only to employ what already exists.

Soon, the schoolchild's project will not be a dog-earred paper, illustrated with photographs from National Geographic, but a disc or tape with computer-generated text and graphics, and video footage taken from a central library. Young children, with their insistence on hearing and seeing the same thing over and over again in the very same way, are already being drawn into a world of learning and entertainment that no previous generation has known.

And so it goes. The sports enthusiast of the future could play the Wimbledon champion—with a holographic tennis ball, a light-sensitive racket, and a pack of king size filter tips close to hand. Strategy rather than exertion will occupy the gifted amateur.

Speculation of this kind becomes a kind of parlour game, which begins with defining the limitations of things as they are now, and ends in plausible ripostes to every imaginable technical obstacle.

We are discussing only what is possible with technology that already exists. The writable disc, flat screen TV, voice and touch sensitive hardware, satellite and cable are all commercial realities. Hardware is steadily getting smaller, better and more versatile. The quest for miniaturisation (currently, through solid state engineering and optical disc technology) is leading to a new generation of compact, integrated systems structurally as well as cosmetically unlike their forebears. With chips in everything from pens to sewing machines, it is not difficult to see the networks growing (or, in the vision of one enthusiast, the infra-red remote control device in the ceiling running the Kenwood mixer).[3] The configurations we envisage do not wait on the development of new tools (although that, too, is certain), but only on the integration of existing ones.

The home work station will serve not simply as an extension of the office desk, but as a clearing house for information of all kinds. If we imagine every office, factory, shop and home, school, library, museum, community centre and gathering place equipped with at least one such terminal, the implications are truly imponderable.

For one thing, the social importance of people's meeting in groups takes on new meaning when it is possible for them to communicate, work, study and play at a distance. Certainly, for an evening's entertainment, a good deal more time and money is now spent on 'staying in' than on 'going out'.[4]

There could also be a paradox between centralisation and de-centralisation. On one hand, resources can be centralised to the point of channelling all data through a single central processor – information may be prepared centrally and disseminated widely. Yet information transfer can so easily be effected between small and farflung installations, that it is equally possible to have a network with no nucleus. There are arguments for both approaches.

The question of security, from one perspective, and of privacy, from the other, becomes impossibly thorny as the telecommunications network grows larger and more diverse. The idea of total innocents breaking into high security information is a topical subject for speculative popular entertainment, but a waking nightmare for institutions with data to protect.

While the ultimate work station may put the accumulated knowledge of civilisation as we know it at our fingertips, the screen you watch might just be watching you. The truly invidious thing about automated information handling is that there really is no knowing what may happen to data once it is in the system.

We concede that the introduction of the kind of networks described here is only likely to happen piecemeal, with the rich leading the poor by a great length. Some nations only just have television, while others employ information gathering systems more frightening than the Cassandras of any previous generation could have imagined.

The pursuit of technology has already opened great fissures between the rich and the poor, both socially and politically. The automated nations stand in much the same position in this century as the manufacturing nations did in the last. There are on the one hand the 'information rich', and on the other those still struggling for literacy and elementary education. Middle-class kids in first world countries are getting computers for Christmas; the material advantages they enjoy are only becoming more marked with their easy access to the tools of new technology.

At the same time, the computer generation is growing up with a different attitude toward information and knowledge than any generation before it. People who learn to reason like computers are following a relentlessly linear model: it is lateral thinking that distinguishes intelligence, and leads to quantum leaps in scientific thinking. It would be comforting to think that the more inventive computer-based games will engage our imaginations and counter this threat to creative thinking.

It will be interesting, too, to see how dependence on the screen as an information source will affect perception of size and dimension. Already, following the model of western technology has demanded conceptual leaps for the people of many non-western cultures. As we commit more and more of our information to video and computer-based media, we are all having to adjust to using the screen as a source of much more than light entertainment.

The consequences on our bodies alone are already a source of genuine concern: it is neither malingering nor hysteria that provokes complaints of backache, headache and myriad other nagging discomforts in people obliged to sit in front of VDUs for the length of the working day.[5] Consequences of another kind are not immediately physically manifest, but may be even more insidious.

Pliny the Elder relates that it was a group of Phoenician traders, sitting by their fires on the beach, who first stumbled upon the secret of making glass when, raking the burning embers, they found lumps of molten glass amidst the hot sand and ashes. The story is technically improbable, as well as anachronistic, but it may contain a germ of truth.

Glass has been known from prehistory, but control of its manufacture eluded artisans for centuries. Vases made in Egypt around 1460 BC are perhaps our oldest examples of free-standing, manufactured glass objects. It was not until the first century BC that glass-blowing was introduced into the Roman world, and not until the 17th century that English craftsmen perfected the manufacture of transparent glass.

On the other hand, barring Biblical references to slime and bitumen, and the history of resins like amber and shellac, the plastics industry as we know it only got underway in the 1930s, with the introduction of thermoplastics like PVC and polystyrene. The inventors of the last century laid the groundwork, with such products as celluloid and a synthetic billiard ball which caused a mild sensation in the Wild West (where one '...billiard saloon proprietor (commented) that occasionally the violent contact of the balls would produce a mild explosion like a percussion guncap. This in itself he did not mind, but each time this happened, 'instantly every man in the room pulled a gun.'[6]) Now, a century after the idea was first explored, and after only fifty years on the market, there are plastics tougher than steel, and plastics that can hold ton weights on a single thread.

There is in science a rule of thumb which, roughly stated, reckons that the sum total of scientific knowledge squares itself every hundred years. It may be the arrogance of our age but, looking back over the sweep of centuries, it does seem that the tempo has been building, accelerando poco a poco, from the earliest stirrings inexorably toward the cacophony which is this century.[7]

A second, briefer, analogy adds an interesting corollary: in 1784, Lunardi's balloon ascent in London was a media event of the first proportion – the city went wild to see a man float into the sky over Moorfields; by 1983, people had visited the moon and photographed the planets, and the world paid more attention to the Princess of Wales than the space shuttle. The tempo is accelerating, and the gap between people and technology – expressed in both interest and understanding – is widening.

During the Industrial Revolution, innovation was manifest in machines which could be understood by anyone who cared to study them. Now, in what may be called the Technological Revolution, innovation starts well beyond the comprehension of most of the topdeck of the proverbial Clapham omnibus. We are surrounded by technology the complexity of which is not worth explaining to a general audience.

Furthermore, scientific information is accruing far too quickly for any one person to assimilate even its basics: the frontiers have been pushed beyond the comprehension of a single human mind. We are not likely to see again a Faraday or a Bacon, a scientist with a command of most contemporary knowledge. In fact, the domain of 'general knowledge' hardly now touches on even the basics of applied science.

This presents a problem which has its roots in illiteracy and its manifestation in a phenomenon which one commentator has dubbed 'cybercrud' – 'putting things over on people using computers'.[8] People take for granted technology which they do not, and indeed may not need to, understand. Whereas we once created myths and fables to explain the mysteries of the natural world, it is technology that we now accept dumbly. Perhaps what we have really lost is the folklore.

Taking these ideas to a plausible end is a parlour game of another sort – the Advanced stage of the game which at Beginners level tries to deduce from what distant and disparate events dramatic moments in history are conceived.

Imagine only one defensible fantasy: What if the screen and its peripherals replace ink and paper as the medium through which we transcribe information? People might forget how to write. Certainly we would remain literate (the future of the written word at least seems assured), and very likely most people would be able to handle a variety of input devices – possibly including keyboards – but the ancient and intricate art of writing, evolved so slowly across time and cultures, could yet be eclipsed. (The preparation of this very book, in fact, could hardly have exhausted a single felt-tip pen, for it is the product of a word processor.)

Consider, too, that we are committing the legacy of our culture to ever more fragile and secretive storage media. The archaeologists of the year 3000 may know as much about the ancient Assyrians as they do about us. The Assyrians, after all, were great propagandists, and accomplished artists, who recorded scenes of battles and feasts, sieges and lion hunts, building works and the sack of cities, carved in meticulous detail in relief upon stone. They left vivid records that have survived intact over the millenia.

Most of our records, on the other hand, are now kept on paper, much of it of such low quality that a newspaper clipping from 1945 looks as fragile as a parchment manuscript of 1066. And, more and more, we are employing media such as film, tape and disc in their various forms. Film is notoriously subject to deterioration: we have already lost nearly as much of the work of the silent film era as we have of the ancient Athenian stage. While tape and disc are themselves fairly resilient, the data recorded on them is highly vulnerable to damage and wear.

Moreover, each of these will yield up its contents only to its own reading device – the human eye can see pictures in the frames of a film strip, but magnetically and digitally recorded media give no hint of what they contain. Just as we try to reconstruct Roman siege machines, so the archaeologists of the future may try to build videodisc players. It is ironic that we are probably the best documented people in the history of civilisation – and that our records are among the most fragile ever produced.

The technological revolution will certainly lead to events far removed from the immediate concerns of this generation. We can see the beginnings of great changes even now.

Without a doubt, interactive video is a technology which can displace workers from certain jobs. It can in part atone for the jobs it usurps by offering people other things to do with their time. But devoting this extraordinary tool to the production of bread and circuses, while it may be a strong temptation, would be a prodigal waste of knowledge, experience, labour and money. Interactive video is a remarkable innovation, and its benefits socially, economically and politically could be many. But who will exploit it, the state or the entrepreneur? And how is it to be used? These are questions which could occupy a book in themselves.

New methods of information handling will certainly continue to affect the pattern of our daily lives. Very likely, too, they will begin to effect changes in our perceptions and our attitudes. Interactive video technology may give us the key to a great treasure house. But we must not lose the distinction between information and knowledge, between simulation and experience. Wisdom, not often mentioned in the annals of new technology, flows from knowledge and experience.

1 Home Entertainment in the 1980s, International Resource Development Inc, Norwalk, Connecticut, 1982.

2 The Interactive Video Disc Software Market, Frost and Sullivan, New York, 1983.

3 Consultant Michael Grove, speaking at The British Universities Film Council forum 'Higher Education in Focus', London, 8 April 1983.

4 For example, see: Campaign, 2 September 1983, p 23: 'The Sudden Death of the Big Night Out'.

5 DeMatteo, Bob, and FitzRandolph, Katie. 'The Hazards of VDTs'. Ontario Public Service Employees Union, Toronto, 1981.

7 As a footnote to the glass and plastics analogy, consider that by '...1958, 75 per cent of the products made by one of the largest glass manufacturers, Corning Glass Works, were unknown fifteen years previously.' (Phillips, Charles John. Glass: Its Industrial Applications. Reinhold, New York, 1960. P 1.)

6 Kaufman, M., The First Century of Plastics, quoted in Plastics Materials (Brydson, J. A. Butterworths, London, 1983. P 1).

8 Theodor Nelson, quoted in Computer Consciousness: Surviving the Automated 80's (Covvey, H. Dominic, and McAlister, Neil Harding. Addison-Wesley, Reading, 1980. P vi.)

GLOSSARY

ABSOLUTE ADDRESS See ADDRESS.

ACCESS

As a noun and a verb, the retrieval of information from a storage medium such as videodisc, videotape, computer disk or computer tape. See RANDOM ACCESS.

ACCESS TIME

The time it takes to find, retrieve and display a piece of recorded information. ccess time is usually measured at its worst, the longest it can take to get piece of information to another. This is generally a matter of seconds on video tape and disc players, and of milli- or micro-seconds in a computer. See RANDOM ACCESS.

ADDRESS

The place in a computer program, or video tape or disc, at which a given piece of information is recorded. This location is identified by a numeric code, similar to a post office box number. It can be described generally (the relative address) or specifically (the absolute address).

ALPHANUMERIC

Alphabetic and numerical: usually, a code or keyboard containing both letters and numerals.

ANALOGUE

The term used to describe information which steadily flows and changes. Most of the information which comes from the natural world is analogue: time, temperature and voltage, for instance. The opposite of analogue is digital.

APPLICATION

The word used to denote the way in which technology is used in practice. Among the most common applications of interactive video, for example, are training and marketing.

APPLICATIONS PROGRAM

A computer program designed to do one specific job—an accounting system for a business office, for instance, or a training course for a large company. (Compare OPERATING SYSTEM)

ARTWORK

The illustrations or graphics prepared for printed work, film or video. Artwork includes not only original sketches and drawings, but also captions, titles, photographs, maps, graphs and charts and all the other still (i.e. non-moving) material which goes into a video or film.

ASCII

American Standard Code for Information Interchange: the system used internationally to code alphabetic, numerical and other symbols into the binary values used in computing.

ASPECT RATIO

The width-to-height ratio of a single frame of film or video. Video has an aspect ratio of four units of width to three units of height, and film a ratio of three to two.

ASSEMBLY

In computing, the conversion of instructions and data written in a computer language or even in everyday terms, into the machine code which the computer understands.

ASYNCHRONOUS

In computing, a system in which the stages in a program are queued so that the completion of one operation initiates the next. (Compare SYNCHRONOUS)

AUDIO HEAD

The head on a videotape machine that records and plays back the audio (sound) track/s recorded along one edge of the videotape. The audio head usually stands to one side of the video head or video head drum.

AUDIO STILL FRAME

A single still image (artwork, say, or a slide) accompanied by commentary, music or sound effects. The soundtrack may be recorded using some version of compressed audio for greater economy.

AUDIO TRACKS

The tracks on a video tape or disc which record the audio (sound) signal that accompanies the video (picture) signal. The audio track usually runs along the edge of the videotape, and beside the video track on a disc. Systems with two separate audio tracks can offer either stereo sound or two independent soundtracks.

AUTHORING

The preparation of a computer program, often using an 'author language' or 'authoring system' that allows people without formal training in computer programming to prepare applications programs for computer-based systems.

AUTHORING LANGUAGE

A high level computer program, itself often based on a computer language like BASIC or Pascal, that facilitates the preparation of computer programs by reducing the number of instructions involved and translating these into a language resembling everyday English.

AUTHORING SYSTEM

A collection of authoring programs that allows users without formal computer programming skills to prepare applications programs, often working in everyday language, and without the painstaking detail of computer programming proper.

BACK UP

In computing, a 'safe' copy of data recorded on diskette or other storage medium.

BANDWIDTH

The range of signal frequencies that a piece of audio or video equipment can encode or decode. Video uses higher frequency signals than audio, and so requires a wider bandwidth.

BAR CODE, BAR CODE WAND

A block of optically-coded parallel lines, read by a scanner or a slender stick which transmits a coded message to a microprocessor or computer. Bar codes are most familiar in retail sales, but also appear in workbooks, labels and other printed material.

BASIC

'Beginner's All-Purpose Symbolic Instruction Code': a popular computer language in microcomputing.

BETAMAX

Sony's domestic videotape cassette format; Betamax uses ½" (12.65 mm) tape in a 6" x 3¾" (155 mm x 95 mm) cassette.

BINARY NOTATION

A counting system which uses only two digits, 0 and 1. Where a decimal sequence runs 0 1 2 3 4... a binary sequence would run 0 1 10 11 100... Binary notation is used to represent numerals, letters, symbols and other data. The binary system is suited to computers because its two values, 0 and 1, can be transmitted by an electric current as pulses as 'weak' and 'strong' or 'off' and 'on', and recorded on a magnetic storage medium as 'unmagnetised' and 'magnetised'.

BIT

A binary digit, either a 0 or a 1: a bit is the smallest unit in computer information-handling. A computer's processing capability is usually reckoned in the number of bits which can be handled at one time: personal computers, for example, commonly offer 8-

or 16-bit microprocessors, while a minicomputer may have a 16- or 32-bit central processor. (Compare BYTE)

BLACK BOX

In interactive technology, the intermediary piece of hardware used to link the computer to peripherals such as a video tape or disc player. The black box often comes as part of a commercial interface package.

BOARD (or CARD)

The card which holds the chips and wiring that control either some essential function of the computer's central processor, or a special feature such as interface with a video player. See INTERFACE.

BUG and DE-BUG

As a noun, flaws in a computer program or system. As a verb, testing for 'bugs' and removing them (i.e., 'de-bugging').

BYTE

'By eight': a set of eight bits (binary digits). It generally takes six to eight bits (one byte) to form a single ASCII character. A computer's storage capacity, or 'memory', is reckoned in kilobytes, or K's: one K is actually 2^{10}, or 1024, bits. A typical personal computer might have an 8-bit central processor and 64K of memory. (Compare BIT)

CABLE

The wires and cords and flexes – the 'spaghetti' – used to plug in and link different pieces of equipment.

CAD/CAM

Computer-aided design and manufacture.

CAI

Computer-assisted (or computer-aided) instruction.

CAL

Computer-assisted (or computer-aided) learning.

CAPACITANCE

Electrical capacity: the ability to store an electric charge. One type of videodisc technology uses variations in capacitance between the disc and a pickup stylus or sensor to transmit recorded video and audio information. Two, incompatible formats are based on this principle. RCA's CED (Capacitance Electronic Disc), SelectaVision, uses 'grooved' discs; JVC's VHD (Video High Density) uses 'grooveless' ones. See CED, VHD.

CAPACITOR

The component of an electronic circuit which stores and releases voltages.

CARD

A computing term used synonymously with 'board'. See INTERFACE.

CASUAL USER

Someone who uses specialised equipment (such as a computer or word processor) only from time to time, and who may, therefore, be less comfortable or adept in using the system than a full-time operator would be.

CATHODE RAY TUBE (CRT)

The picture tube of a TV set or video monitor, or of a computer's VDU (visual display unit). The CRT is a vacuum tube (usually made of glass) containing, essentially, an electron gun and a luminescent screen. The video picture, transmitted as a series of electrical signals, is reconstituted as light and colour within the CRT.

CAV

Constant Angular Velocity, or 'active play' videodiscs. CAV discs spin at a constant speed of 1500 rpm (PAL) or 1800 rpm (NTSC), and assign a variable track length to each frame. This is the equivalent in both systems to a rate of one frame per revolution, which means that individual frames can be identified and retrieved quickly and easily–the 'rapid random access' which is a basic requirement of interactive video. (Compare CLV)

CAVIS

Computer Aided Video Instruction System: the interactive video delivery system developed by Scicon, a subsidiary of British Petroleum. CAVIS is an expensive system aimed at the corporate in-house training market. CAVIS comprises a Panasonic VHS player, a 64K microcomputer, a dot matrix printer, a keyboard for the instructor and a keypad for the student. It can generate text and graphics in colour.

CBT

Computer-based training.

CED

Capacitance Electronic Disc: RCA's videodisc format, 'SelectaVision'. Information is encoded in a series of shallow pits running along a spiral groove. The disc is read by a metal electrode mounted in a diamond or sapphire stylus. Variations in capacitance between the disc and the electrode are converted into audio and video signals. An NTSC disc runs at 450 rpm, with four complete frames per revolution. The disc is made of PVC, and is housed in a case which protects it from pollution, damage and wear.

CENTRAL PROCESSOR

The 'brain' of the computer, to which all the various parts of the computer are ultimately linked, and where all the 'processing' of information – instructions, calculations, data manipulation – takes places. Also called the CPU (Central Processing Unit).

CHAPTER

One independent, self-contained segment of a computer program or interactive video programme.

CHAPTER STOP

A code embedded in an interactive videodisc to signal the break between two separate chapters, so the programme can start at the beginning and stop at the end of any one.

CHIP

A device in which microscopic electronic circuitry (such as that forming a transistor or integrated circuit) is printed photographically on the surface of a tiny piece of semiconductor material (usually, crystalline silicon).

CHROMINANCE SIGNAL

A signal which carries information about the colour of a TV picture. The luminance signal carries information about the picture itself, in black and white.

CLV

Constant Linear Velocity, or 'long play' videodiscs. CLV discs assign a fixed track length to each frame and spin at a speed which gradually decreases as the disc plays. This increases the total playing time, but restricts access to terms of chronological time (minutes and seconds). CLV discs are a good medium for entertainment, but of little use interactively.
(Compare CAV)

COLOUR STANDARD

The way in which a colour picture is composed and transmitted. There are three distinct, incompatible colour standards: PAL, NTSC and SECAM, plus some minor variations within the PAL and SECAM systems. (There are six monochrome – black and white – standards, as well.)

COMPATIBILITY

The facility to use elements of different systems either in combination or interchangeably. The various videotape systems – U-matic, VHS, Betamax, V2000 - are incompatible: a tape from one system won't even fit a machine from another. Laser disc systems are essentially compatible with one another, but the three basic disc systems – laser discs, VHD and CED – are mutually incompatible. The world's two principal video production standards, PAL and NTSC, are incompatible - totally different.

COMPRESSED AUDIO

A system of recording and transmitting audio information in highly compact form by encoding and decoding conventional audio signals digitally, or by converting audio signals into video signals for more compact storage. This greatly increases the versatility of the video tape or disc.

COMPUTER

Essentially, an electronic device which can input, store, process and output information of different kinds. Hardware describes the actual physical components, software the computer programs. Analogue computers interpret data as variable physical quantities (such as voltage); the more common digital computers employ binary notation to translate all data into fixed, numeric values. Dedicated computers can only do one specific job;
universal or multi-purpose use different software programs to do a variety of jobs. The hardware comprising a typical computer system includes a central processor (CPU), a screen (or visual display unit, VDU), a keyboard and/or some other control device, a printer and/or plotter.

COMPUTER-GENERATED

Usually, text and graphics created, stored and produced entirely by a computer - either the elaborate equipment in a professional editing suite, or the external computer used in a Level 3 interactive video configuration.

COMPUTER GRAPHICS

Graphics–line drawings, pictures, charts, graphs, and so on–created through a computer.

CONSUMER See DOMESTIC.

CORE STORAGE

The main memory store of a computer. Properly, the term refers to ferrite rings, or cores, used for magnetic storage in older computers. Whilst these have been superceded by integrated circuits or chips, which are faster and more compact, the term is still current.

COURSEWARE

That part of an interactive video training or teaching course comprising the video programme (on disc or tape), and its complementary computer program(s), including those generating text and/or graphics.

CPU See CENTRAL PROCESSOR.

CURSOR

The flashing shape on the computer screen which indicates where information may be entered next.

CVC (COMPACT VIDEO CASSETTE)

The ¼" (6.3 mm) videotape format, which employs a cassette the size of a conventional audio cassette; sometimes described as '8 mm video'.

DATABASE

A source or file of information which users may employ or amend as part of a larger project or program.

DE-BUG See BUG.

DEDICATED

In computing, a system which performs one special job (such as an airplane's autopilot, or certain interactive video delivery systems), and cannot be used for any other. In the design of computer and other keyboards, a global key which controls some particular function of the machine, doing the same thing every time it is pressed. (Compare UNIVERSAL)

DELIVERY SYSTEM

In interactive video, the set of video and computer equipment actually used to deliver the interactive video programme. A delivery system may comprise as little as a videodisc player with onboard microprocessor, a monitor and a keypad, or may extend to an external computer, two or more monitors, and a variety of peripherals.

DIGITAL

In computing terms, data which is generated as or translated into a pattern of discrete, fixed values, such as digits or other concrete characters. The opposite of digital is analogue. Digital information usually refers to computer- based technology, and employs binary notation.

DISC, DISK, DISKETTE

In audio, video and computer technology, a flat circular plate, usually composed chiefly of plastic, which can be used to record and replay various types of information, both analogue and digital. An audio record is made of vinyl, with sound signals electronically encoded along a spiral track which undulates from side to side; it spins at 33, 45 or 78 rpm. A videodisc is usually composed of layered plastics, with video, audio and programming information encoded as a series of shallow pits along a spiral track; different formats spin at speeds between 450 and 1800 rpm. The 'floppy' disk (or diskette) used by microcomputers is a thin, flexible, magentically-coated plastic disc, housed in a firm square plastic envelope; standard sizes are 5" and 8" in diameter. 'Disc' is a common spelling in Britain and Europe, and 'disk' in the USA; a nice distinction uses 'disc' for video and 'disk' for computer technology.

DISK DRIVE See DRIVE.

DOCUMENTATION

The papers which record how an interactive video programme—and particularly its computer program—was designed.

DOMESTIC

The term used synonymously with 'consumer' to describe video tape and disc players designed principally to cater to the home market. Equipment designed for professional, non-broadcast work is usually described as 'industrial'.

DRAW (DIRECT READ AFTER WRITE)

A field of recording technology in which discs are checked for accuracy (by being 'read') immediately after they are recorded (that is, 'written').

DRIVE

That part of a computer-based system such as a personal computer or a word processor into which the floppy disks are inserted when they are being used to input, process or output information.

DUB

Usually, to replicate a video or audio recording on tape or disc. Sometimes, to lay a new or additional audio track onto an audio or video tape recording.

EDIT

In video, to create the master tape of a video programme, usually from a variety of source media, or, specifically, to link one piece of audio or video tape to another (a bad edit is an imperfect join which can be clearly detected when the tape is replayed).

EDITING SUITE

The place in which editing actually happens. There are professional editing facilities which usually offer suites containing different configurations of equipment, which can be rented by the hour. Many organisations which use video internally have their own production and editing facilities.

ELECTRONIC

Relating to electronics, an applied science involving physics and technology and concerned with electrons, those negatively-charged particles, found in all atoms, which carry electricity in solid matter. Electronic equipment is based on solid state engineering (thermionic valves, transistors, integrated circuits, chips and so on), vacuum or gas tubes (such as cathode ray tubes), and similar devices designed to control the motion of electrons.

EXECUTE

In computing, to carry out commands.

FELIX

The interactive video delivery system developed by Felix Learning Systems for the training market. There are three levels of work station, built around a U-matic video tape player and monitor, a microcomputer (48K for students and instructors, 64K for programmers), a keyboard, and a printer. Users cannot make their own programmes, but can buy programmes off the shelf, or commission them.

FIELD

A scan of one half of the lines which comprise a single, complete frame on the video screen. Two interlaced fields make up one frame. PAL standard systems employ a 625 line screen, so that there are 312.5 lines in a field; NTSC systems use a 525 line screen, so there are 262.5 lines in a field.

FIELD DOMINANCE

The field, even or odd, on which any given frame begins. Some editing equipment always starts on odd-numbered fields, some on even. It is important to establish which before starting the edit, to avoid 'flicker' between the fields.

FIELD STANDARD

The video production standard which effectively describes the running speed of the video programme. There are two interlaced fields in one complete frame.
The field standard tends to relate to the mains power supply frequency. Where the mains supply is 50 Hz, broadcast television (and, hence, non-broadcast video) runs at 50 fields (or 25 frames) a second; this is the standard employed by PAL and SECAM systems. Where the power supply is 60 Hz, the field standard is 60 fields (or 30 frames) per second; this is the standard employed by the NTSC systems.

FLICKER

The subtle yet perceptible flashes of light between frames of moving footage, or between imperfectly matched fields of a freeze frame. Flicker is only detectable up to about 45 Hz, so the observer is not aware of flicker in an electric lightbulb running off a mains power supply of 60 Hz, or a television broadcast running at 50 or 60 fields a second.

FLOPPY, FLOPPY DISK, FLOPPY DISKETTE See DISC.

FLOWCHART

A diagram representing a computer program in graphic form—usually as a series of variously-shaped boxes joined, where appropriate, by straight lines. There is not yet a standard graphic language for interactive video flowcharting, although in computer flowcharting many basic elements are identified with specific shapes on the flowchart.

FONT or FOUNT

In printing and computer-generated character generators, the range of characters and special symbols available in one style and size of type.

FRAME

A single, complete picture in a video or film recording. A video frame comprises two interlaced fields. Film runs at the rate of 24 frames a second; video at 30 frames a second in NTSC standard systems, 25 frames a second in PAL and SECAM systems.

FRAME-ORIENTED

An authoring system in which the author works directly on designing the screens which comprise the basis of the finished programme. (Compare LINE-ORIENTED)

FREEZE FRAME

A single frame from a strip of moving footage held motionless on the screen of the video monitor. Unlike a still frame, a freeze frame is a not a picture originally shot to appear on its own, but is one frame taken from a longer moving sequence.

FREQUENCY

The number of times a signal vibrates within a given time; usually measured as Hz (Hertz) or cycles per second.

GENERATION

In hardware, the level of available technology, the first generation usually being identified with the first production model. In storage media, the number of times a reproduction is removed from its original source.

GENERATOR

In computer-based technology, a device or system which facilitates some job such as designing graphics, text or programs.

GLOBAL KEY

A key dedicated to some specific function, which performs that function and no other, every time it is pressed.

GRAPHICS

The term used to cover the artwork incorporated into a video programme: sketches, line drawings, charts and so on.

GRAPHICS TABLE or TABLET

A sensitive board which acts as a canvas through which computer-generated graphics can be designed. A handheld input device such as a light pen is used to draw freehand, to block out geometric shapes and to transmit instructions.

GUARD TRACKS or GUARD BANDS

The protective bands sometimes inserted between tracks of recorded material on audio or video tape to prevent 'cross- talk' between two unrelated tracks.

HARD COPY

A printout of information – a computer program, for example, or student scores - from a computer on to paper.

HARD DISK

An inflexible magnetic disk with greater storage capacity than a floppy diskette. A hard disk is sometimes sealed within a computer to provide a large (often temporary) memory into which data from several of the smaller, handier floppy disks can be loaded before running a long or complex program. The Winchester disk is one such hard disk.

HARDWARE

The physical equipment which actually makes up a computer or interactive video delivery system: the tangible side of the operation. The complement of hardware is software.

HELICAL SCAN

The videotape system which employs two video heads, mounted on opposite sides of a revolving drum. The video head drum spins at the rate of one frame per revolution (25 rps for PAL, 30 rps for NTSC), so each head scans one field (half of one frame) per revolution. Helical scan achieves the high tape speeds needed for video recording by moving both the tape and the video heads: a Betamax tape with a linear speed of 1.9 cm/s has a 'writing speed', relative to the video heads, of 6.6 m/s. The name derives from the helical shape described by the tape as it winds around the video head drum.

HIGH LEVEL See LANGUAGE.

Hz

Hertz, the standard unit of frequency: one Hz is equal to one cycle per second. One kilohertz (kHz) = 10^3 cycles per second, one megahertz (MHz) = 10^6 cycles per second. (Heinrich Hertz, German physicist, d. 1894)

INDUSTRIAL

The term used to describe equipment designed for professional, non-broadcast applications, such as corporate communications, training, marketing and education.

INFORMATION BLOCK

One segment in an information programme, comprising some combination of moving footage, stills and computer-generated text and graphics and excluding menus, tests and the like.

INFORMATION PROGRAMMER

The person on an interactive video production team who translates the work of the instructional designer into a computer program.

IN-HOUSE

Entirely within the given company or organisation, using its resources, facilities and expertise.

INPUT

The term generally used, as verb and noun, to describe the transfer of information from any source into a computer.

INSTRUCTIONAL DESIGNER

The person in the interactive video team who, given the aims and objectives of the project, analyses the content of the programme and arranges that material in a way that can then be used by the information programmer.

INTEGRATED CIRCUIT (IC)

A complete electronic circuit (the path along which an electric current travels) chemically produced on the surface of a single chip of semiconductor material (usually, a tiny piece of crystalline silicone).

INTERACTIVE

Involving the active participation of the user in the directing the flow of the computer or video programme. The opposite of interactive is linear.

INTERACTIVE VIDEO

The convergence of video and computer technology: a video programme and a computer program running in tandem under the control of the person in front of the screen. In interactive video, the user's actions, choices and decisions genuinely affect the way in which the programme unfolds.

INTERFACE

As a general term, the link between two pieces of disparate equipment, usually the computer's central processor and its peripherals. As hardware, the circuitry that effects this.

INTERLACE

The pattern described by the two separate field scans when they join to form a complete video frame. As the video picture is transmitted, the first field picks up even- numbered scan lines, the second, odd-numbered ones. The two lace together to form a single complete frame.

JOYSTICK

A remote-control device, similar in appearance to the gearstick of a car, popularly used in video and arcade games, and in some interactive video applications.

JUDDER

The shaky or shivering effect which appears between mismatched fields in a still or freeze frame.

K (Kbyte, Kb)

The abbreviation for kilobyte, actually, 2^{10}, or 1024 bytes, that being the nearest binary notation comes to 1000. A computer's size is often reckoned by the number of kilobytes' memory it offers.

KEYBOARD

A board containing alphanumeric and other keys used to create text and convey instructions to a computer. The computer keyboard usually contains, in addition to the familiar characters and symbols of a typewriter keyboard, keys dedicated to specific computing functions.

KEYPAD

A remote control device, such as the remote control unit of a television, containing a selection of keys dedicated to specific functions and, usually, a set of numeric keys.

KEYWORDS

Words which the computer recognises. Keywords (themselves written in a high level computer language) are often used as shortcuts in computer programming, and in specific applications such as training, where 'keyword' tests help evaluate the user's comprehension of the material.

LANGUAGE

A medium of communication between the computer programmer and the computer. Among the popular computer languages are COBOL (Common Business Oriented Language), BASIC (Beginner's All-purpose Symbolic Instruction Code), Pascal, the mathematical and scientific languages ALGOL and FORTRAN, and the hybrid PL/1 (Programming Language 1). A 'high level' language (such as an authoring language) closely resembles everyday language, while a 'low level' language is more closely related to the computer's own binary machine code.

LASER

The acronym of Light Amplification by Stimulated Emission of Radiation. An atom at its most stable energy state is said to be at a ground state; at higher energy levels, it is said to be 'excited'. A laser exploits the energy of excited atoms to produce an intense beam of electromagnetic radiation – light.

The laser is contained in a transparent cylinder with a reflective surface at one end, and a partially-reflective surface at the other. The stimulated waves of atoms travel up and down the tube, and some pass through the semi-reflective end as light beams. That used in videodisc technology is currently usually a helium neon laser, built from a mixture of those two inert gases.

LASER DISC

The name popularly used to describe the reflective optical videodisc. See REFLECTIVE OPTICAL VIDEODISC.

LASERVISION

The name of Philips' reflective optical videodisc – or laser disc – system.

LEVEL 0

The bottom of the scale designed by the Nebraska Videodisc Design/Production Group to describe interactivity in videodisc players. Level 0 represents domestic-standard players which have no potential for interactivity.

LEVEL 1

The first practical level of the Nebraska scale. Level 1 represents the basic features expected even on domestic equipment: remote control, the 'search' facility, freeze frame, forward and reverse motion, and quick scan, slow motion and step frame replay.

LEVEL 2

The mid-point on the Nebraska scale. Level 2 player is one which, using only its own onboard microprocessor, can offer multiple choice, the branching facility and score-keeping.

LEVEL 3

Effectively the top of the Nebraska scale. Level 3 represents a video player, industrial or domestic, linked to an external computer (mainframe, mini or micro). Level 3 offers by far the greatest versatility of any interactive configuration.

LEVEL 4

In a practical interpretation, a complete workstation comprising video and computer equipment and furniture. Speculatively, the peak of the Nebraska scale.
Level 4 still represents conjecture into the future of interactive technology, rather than actual applications.

LIGHT PEN

A remote control device which allows the user to write or draw on the screen of a cathode ray tube with an extremely sensitive photo-electric 'pen'. Light pens can be used to 'read' the surface of the screen, to input information or to modify recorded data, and to interact with a teaching or training programme.

LINE-ORIENTED

An authoring system in which the author designs a programme (including screens of text or graphics) as a long string of commands to the computer. (Compare FRAME-ORIENTED)

LINE STANDARD

The video production standard which describes the number of 'scan lines' used by a given television system to make up one screen. Line standards from 405 to 819 have been used; NTSC now uses a 525 line screen, PAL and SECAM, 625 lines.

LINEAR

The opposite of interactive: a video programme which plays straight through from start to finish without interruption.

LOCATION

In computing, the place in the store where data can be recorded, usually discussed in terms of its address. In video and film production, a location is a place at which material is shot in an environment which represents the 'real' world – as least in comparison with a studio.

LOW LEVEL See LANGUAGE.

LUMINANCE

The broadcast signal which carries information about the composition of the picture on the TV screen. This information is carried in black and white: the chrominance signal adds the colour.

MACHINE CODE

The binary code (that is, the pattern of 0's and 1's) which is the computer's native language. High level computer languages, which more closely resemble our everyday language, are ultimately converted back into machine code within the computer.

MAGNETIC DISK

A flat circular plastic plate, magnetically-coated, which can record and replay both digital and analogue signals. A rigid magnetic videodisc has been used for instant replay in television sports broadcasting since the late 'sixties, but the most common magnetic disk is the computer's 'floppy'. See DISK.

MAGNETIC TAPE

A thin, strong, non-elastic plastic tape coated with a ferromagnetic emulsion, which can record, store and play back information of various kinds. Audio tape records sound, videotape sound and pictures (as well as electrical signals used in editing and in interactive video applications). Magnetic tape is also used as a computer information storage device.

MAINFRAME

Originally, the main framework of a computer's central processing unit; subsequently, the CPU itself; popularly, the largest of computers, both in size and capacity, and the most expensive. (Compare MINICOMPUTER, MICROCOMPUTER)

MASTERING

A stage in the production of a videodisc, in which the master disc (from which all subsequent discs will be ultimately be pressed) is cut.

MEMORY

The place in which computer-based equipment stores recorded information, either permanently or temporarily.

MENU

In computer-based technology, a page on the screen which lists options available to people using the system for some purpose – such as preparing an interactive video programme, or working with one.

MENU-DRIVEN

A programme which is built around a series of menus, or tables of contents, which guide users through through the options available to them.

MICROCHIP See CHIP.

MICROCOMPUTER

Currently, physically the smallest of computers. Also known as a 'home' or 'personal' computer.

MICROPROCESSOR

A device which offers limited computer power and storage in equipment which is not essentially computer-based – such as a videodisc player.

MINICOMPUTER

The middle range of computers, physically at least. Many modern micros offer as much power and versatility as minis used to do.

MODEM

A device for converting information for the computer into signals which can be transmitted over ordinary telephone lines. From MOdulate/DEModulate.

MONITOR

In video, an electronic device, similar to a television, which receives and displays a non-broadcast video signal sent across wires within a closed circuit (from, say, a video tape or disc player), but which cannot intercept a broadcast signal. In computing, another name for the computer screen. (Compare RECEIVER)

MOUSE

A remote control device, rather like a toy mouse on wheels, which can be guided by hand around a tablet or other sensitive surface to direct a cursor on the computer screen, often in games-playing or the design of computer graphics.

NEBRASKA SCALE

The scale devised by the Nebraska Videodisc Design/Production Group to describe interactivity in videodisc players. See LEVEL.

NOISE

The audible interference a signal picks up every time it passes through a piece of electronic circuitry.

NTSC

The colour standard established by the American National Television Standards Committee. NTSC is used as a general term to describe video systems which employ the whole American broadcast standard, with a 525-line screen, and a running speed of 60 fields/30 frames per second.

OPERATING SYSTEM

The set of programs which control the running of a computer and its peripherals, and dictate what software can be used for individual APPLICATIONS PROGRAMS.

OPTICAL VIDEODISC:

See REFLECTIVE OPTICAL VIDEODISC.

OUTPUT

Material generated by a computer from its memory for display on a screen, or transfer to some hard medium such as paper through a printer or plotter.
OVERLAY

The facility to superimpose computer-generated text and graphics over a video picture, moving or still.

PA

In video, the production assistant, an individual whose duties on different projects and in different companies may range from making the tea to running the production team.

PACKAGE

The term used to describe a set of compatible, interlinked equipment designed to make up a complete delivery system, or a set of computer programs needed to handle one specific job, such as video/computer interface.

PADDLE

A manual control device adopted by interactive video from computer and video games.

PAL

Phase Alternating (or Alternate, or Alternation) Line: the most widely applied of the three colour standards. PAL was developed in West Germany, and is employed in the UK and across most of Europe, Africa, Australasia and South America.

PARALLEL

In computing, the approach which sends all the bits in a byte at one time, 'abreast' as it were, between the central processor and the computer's peripherals.

PASCAL

The computer language which recalls Blaise Pascal (1623-62), the French philosopher, physicist and mathematician who, among other accomplishments, invented the first mechanical calculator.

PERIPHERALS

Properly, peripheral units: equipment controlled by the computer, but physically independent of it; for example, the computer's keyboard, printer or plotter, and the video equipment of an interactive video system.

PHOTODIODE

A device used in an industrial standard laser disc player to translate variations in the light reflected off the pitted surface of the disc into the electronic signals which comprise the audio, video and control tracks of the programme.

PIXEL

An abbreviation of 'picture element'—one of the thousands of points of light and colour which make up a computer screen. There are currently up to around 1500 x 1200 pixels to a screen.

PLOTTER

A computer output device which displays information in graphic form, usually by plotting a graph on paper.

POINT-OF-SALE/POINT-OF-PURCHASE

The term used to describe interactive video units set up in stores and shopping precincts to demonstrate products, sell goods, advertise services, draw attention to sales and special offers, provide community information and so forth to passers-by. 'Point-of-sale' (POS) is preferred in the UK, 'point-of-purchase' (POP), in the US and Canada.

PORT

The socket at which cables connecting the computer and its peripherals are attached.

POST-PRODUCTION

The stage in the preparation of a film or video programme after the original footage has been shot. Post-production includes the editing stage and the preparation of the tape or film for reproduction as an interactive tape or disc.

PR 7820

A popular early NTSC-standard, industrial quality videodisc player, launched by DiscoVision Associates (DVA) and manufactured by Pioneer after their takeover of DVA; followed by the PR 8210 and other, later models.

PRE-MASTERING

The stage in the production of a videodisc when the master tape is checked and prepared for transfer onto the master disc from which all subsequent discs will be pressed.

PRE-PRODUCTION

All that part of the production schedule leading up to the actual shooting of material on video or film.

PRINTER

A computer output device which records information by printing on paper under the direction of the computer. A daisywheel printer uses a small disc, shaped like a rimless wheel, with a character at the end of each spoke. A dot matrix printer represents textual and graphic information as patterns of small dots.

PROCESSING

In computing, the manipulation of data from one state to another, usually at the request of an operator or user. In cine or still film, the photographic development of film footage.

PRODUCTION

In video terms, that stage in the job when video or film footage is actually shot. (Compare PRE-PRODUCTION and POST- PRODUCTION)

PROFESSIONAL

Synonymous with 'industrial' as a description of video equipment.

PROGRAM and PROGRAMME

As a useful distinction, 'program' refers specifically to computer software, and 'programme' to material on video tape or disc, or interactive video systems generally.

PULSE, PULSE CODE

A signal which, when recorded on every frame of a videotape, facilitates editing and access by making individual frames easier to identify.

QUADRUPLEX

The professional videotape system, developed by Ampex, which employs four video heads mounted 90° apart on a drum which spins at 240 rps (NTSC) or 250 rps (PAL). Quad uses broadcast quality 2" (50 mm) videotape.

RAM

Random access memory—that part of a computer's memory which can both read (use and display) and write (record) information, and which can be updated or amended by the programmer or user. This is the largest part of a computer's memory, which it uses in its day-to-day work.
(Compare ROM)

RASTER

The pattern of scan lines on a conventional video receiver or monitor. The word derives from the German for 'rake'.

READ

To transfer information from one (usually permanent) storage medium to another (often temporary) one: for example, to transfer data from a floppy disk to a computer screen, or pictures and sound from a videotape to a monitor.

REAL ESTATE

In video technology, the available space on a disc or tape.

RECEIVER

An electronic device, such as a television or radio, capable of receiving and display a broadcast signal. (Compare MONITOR)

REFLECTIVE OPTICAL VIDEODISC

The format that uses lasers to write (record) and read (play back) a videodisc. A writing laser turns video, audio and control signals recorded on videotape into a pattern of shallow pits along a spiral track on a glass master disc. Copies are pressed in plastic with metal 'stampers' molded on the glass master. These copies are covered with reflective aluminium film and sandwiched between layers of translucent plastic. In play, the disc spins at 1500 (PAL) or 1800 (NTSC) rpm. Light from the reading laser bounces off the reflective surface through a photosensitive diode that converts variations in reflected light into electrical signals from which video and audio signals derive. Optical disc technology, of which videodisc is a part, describes the whole field of (computer) information storage on disc. Optical discs can store a variety of information including computer-generated data and digitially-encoded video and audio signals.

RESOLUTION

The fineness of the detail in a video or computer display screen, measured as a number of discrete elements – dots in a video screen, pixels on a VDU. The higher the number, the higher the resolution and the better the picture.

RGB

Red-green-blue – a high quality colour screen used with many computers, and increasingly with video systems as well.

REMOTE CONTROL

Control of a computer program or interactive video programme through an electronic device independent of the computer or video player. This can be effected through a simple keypad or periperhals such as the computer keyboard, a touch- sensitive screen, the joysticks and paddles of computer games, or any object suitably wired to communicate with the user and the system.

RESPONDER

Sony's interactive video delivery system, designed for low budget training applications. The system comprises Sony's U-matic tape recorder/player, a receiver/monitor, a student unit with ten numeric and three dedicated keys, a printer, a 'cue programmer' which allows the instructor to programme tapes, and a 'dub converter' for duplicating tapes.

ROM

Read-only memory, the smaller part of a computer's memory, in which essential operating information is recorded in a form which can be recalled and used (read) but not amended or erased (written). (Compare RAM)

RS-232C

A standard serial interface between a computer and its peripherals. Various peripherals (including some videodisc players) equipped with an RS-232C computer port can be plugged directly into a compatible computer. The RS-232C is increasingly becoming a standard feature of computers and their peripherals.

SAFE AREA

That area in the centre of a video frame which is sure to be displayed on all receivers and monitors. Televisions and monitors made at different times and by different companies are slightly different in size and shape, and the outer edge of the video frame (about 10% of the total picture) is not reproduced in the same way on all sets.

SATURATED COLOURS

Strong, bright colours – particularly, reds and oranges – which do not reproduce well on video, but tend to 'saturate' the screen with colour, or 'bleed' around the edges, producing a garish, unclear image.

SCAN

In basic television and video transmission, the rapid journey of the scanning spot back and forth across the scan lines on the inside of the screen. In interactive video technology, the facility to move quickly backwards or forwards through the programme.

SCAN LINES

The parallel lines sloping across the video screen from upper left to lower right, along which the scanning spot travels in picking up and laying down the video information which makes up the picture on the TV screen or monitor. PAL standard systems employ 625 lines in a screen, NTSC standard systems used 525.

SCANNING SPOT

The intense beam generated by the electron gun in a cathode ray tube (picture tube), which travels rapidly back and forth across the screen, picking up and laying down the information which makes up the picture on the screen.

SEARCH

The facility in interactive video systems to request a specific frame, identified by its unique sequential reference number, and then to instruct the player to move directly to that frame, forwards or backwards, from any other point on the same side of the disc or tape.

SECAM

'Séquential couleur à mémoire' (sequential colour with memory), the colour standard developed in France, and subsequently adopted by the USSR and its satellite states, and in some parts of the Middle East and North Africa. It involves sending the three primary colour signals sequentially, rather than nearly simultaneously (as the NTSC and PAL systems do).

SELECTAVISION

The name under which RCA markets its CED (Capacitance Electronic Disc) video disc in the United States.

SEMICONDUCTOR

An electrical conductor which allows a small current to pass in one direction. Semiconductors are used to carry the pattern of electrical signals, off/on, weak/strong, which represent the 0's and 1's of computing's binary notation.

SERIAL

In computing, the approach which sends all the bits in a byte one at a time, 'single file', between the central processor and the computer's peripherals.

SERVO-CONTROL

A device which converts a small mechanical force into a larger one, particularly in a control mechanism.

SIGNAL-TO-NOISE RATIO

The strength of a video and/or audio signal in relation to the interference, or 'noise', it has picked up passing through electronic circuitry. The higher the number, the better the quality of the signal.

SLOW MOTION

The facility to move backwards or forwards through a video sequence at an exaggeratedly slow speed.

SOFTWARE

The programs which actually make the computer run. Software represents the intellectual side of computer-based technology, hardware the physical side.

STANDALONE

Equipment such as a computer terminal or interactive video system, which is independent of any larger network.

STEP FRAME

The facility to move through a video sequence frame by frame, forward or backward, either automatically or using a remote control device. This can be used to examine a sequence of moving footage in close detail, or to employ a set of stills which have been recorded as single, static frames.

STILL FRAME

A graphic of any kind which is presented as a single, static image rather than as moving footage. The economical storage of still frames is one of the strengths of the videodisc.

TeD, or TELEDEC

An early videodisc system, developed jointly by Germany's Telefunken and Britain's Decca. It employed a flexible plastic foil disc, read by a prow-shaped stylus. However, the first discs were only ten minutes long, and picture quality was poor; the system appeared only briefly on the European market in 1975.

TELECINE

Equipment used to make a videotape copy of a piece of film footage, or to copy transparencies (colour positives) from photographic slide film onto videotape.

TELETEXT

The generic term used to describe systems which use broadcast television signals to transmit digitally-coded information. In Britain, the BBC's public service is called Ceefax, and the IBA's, Oracle. Other systems include the Canadian Telidon and the French Antiope.

TERMINAL

The point of communication between the user and a computer- based information system, through which information can be input and output. In computing, terminals are often remote work stations connected to a central processing unit. A typical computer terminal comprises a visual display unit (VDU, or screen), a keyboard, and one or more disk drives.

THREE/TWO PULLDOWN

The method used to reconcile film (which runs at 24 frames a second) to NTSC-standard video (which runs at 30 frames a second). The equivalent of six extra frames, or twelve extra fields, is needed every second: so, when the film is transferred to videotape, on every second frame, one field is recorded twice - adding twelve fields in all.

TOUCH SENSITIVE SCREEN

In computer-based technology, a screen which acts as a control device under the user's touch. Basic functions are executed by touching or stroking certain parts of the screen, and specific responses made by touching appropriate words, messages or pictures as they appear.

TRANSISTOR

A semiconductor device which can amplify signals passing through it.

TRANSMISSIVE OPTICAL DISC

A videodisc format developed by the French telecommuncations company Thomson-CSF. The disc is read by a laser, as are the more familiar reflective optical discs; however, it is translucent, so the reading laser passes straight through the disc, which can thus be read on both sides consecutively simply by re-positioning the laser beam.

TURTLE

A remote control device, similar to a mouse, but larger. It can be guided by hand over a sensitive surface such as a graphics table, its movements directly control those of the cursor on the computer screen. (Remotely controlled devices in Seymour Papert's LOGO computer language are also called turtles, but we are not considering these here.)

UNIVERSAL

In computing, a multi-purpose machine which can address a number of different tasks, given the appropriate software. (Compare DEDICATED)

V2000 (VIDEO 2000) See VCC (VIDEO COMPACT CASSETTE).

VCC (VIDEO COMPACT CASSETTE)

Philips' videotape cassette format, the V2000 (Video 2000), which employs the ½" (12.65 mm) tape of other domestic systems, but writes over only half the width at one time, to produce two ¼" tracks on one ½" tape.

VCR (VIDEO CASSETTE RECORDER)

Generically, a videotape recorder which handles videotape presented in cassettes. Specifically, the name of Philips' original videotape cassette format, the first true video tape recorder on the consumer market. (Compare VCC, VTR)

VDU

Visual display unit – the screen of a computer terminal.

VHD

Video (or Very) High Density (or Home Disc) – JVC's capacitance disc format.
The VHD system employs a 'grooveless' capacitance disc, in which a sensor scans the surface of the disc, following a control track which runs along side the track containing video and audio information. VHD is the smallest of the disc systems, the first to offer high-density sound with still frames, and currently the only one to handle NTSC and PAL on one player.

VHS (VIDEO HOME SYSTEM)

JVC's domestic videotape format. VHS employs ½" (12.65 mm) videotape in a 7½" x 4" (190 mm x 105 mm) cassette.

VIDEO

A system of recording and transmitting information which is primarily visual, by translating moving or still images into electrical signals. These signals can be broadcast (live or pre-recorded), using high frequency carrier waves, or sent through cables on a closed circuit. 'Video' properly refers only to the picture, but as a generic term usually embraces audio and other signals which are a part of a complete programme. Video now encompasses not only broadcast television, but many non-broadcast applications, including corporate communications, marketing, home entertainment, games, teletext, surveillance and security systems, and even the visual display units (VDUs) of computer-based technology.

VIDEO BILLBOARD

An extra-large screen onto which images from video tape or disc, and/or computer-generated graphics, can be projected.

VIEWDATA or VIDEOTEX

The generic term used to describe various systems which use telephone lines or television signals to transmit digitally- coded information. In Britain, the system which uses public broadcast television is called teletext; the BBC's service is called Ceefax, and the IBA's, Oracle. British Telecom's service, Prestel, uses the telephone lines. Other broadcast television systems include the Canadian Telidon and the French Antiope.

VIDEODISC

A generic term describing a medium of video information storage which uses thin circular plates, usually primarily composed of translucent plastic, on which video, audio and various control signals are encoded, usually along a spiral track. Optical disc systems use a laser beam to read the surface of the disc;
they are so far divided between reflective and transmissive systems. Capacitance systems employ a sensor or stylus; they are divided between 'grooved' and 'grooveless' disc systems. A magnetic videodisc is used in broadcast television, and for information storage.

VIDEO HEAD or VIDEO HEAD DRUM

The drum within a videotape player which reads video signals recorded on the tape.

VTR (VIDEO TAPE RECORDER)

Properly, a videotape recorder which handles videotape presented on open reels, rather than cassettes. Popularly, any machine which records and plays back videotape. (Compare VCR)

WRITE

To transcribe recorded data from one place to another, or from one medium to another; information from the computer is written to a disk, rather than on a disk.

WRITING SPEED

The speed at which the video heads on quadruplex or helical scan video recorders revolve in relation to the videotape passing across the video head drum. For instance, a VHS tape with a linear speed of 2.34 cm/s has a writing speed of 4.85 m/s.

X-Y CO-ORDINATES

Points on a plane which has been even divided down (Y) and across (X) on a pre-determined scale. X-Y co-ordinates can be used to plot a drawing or graph on a computer screen.

SOME MANUFACTURERS OF VIDEO EQUIPMENT AND VIDEODISCS

HITACHI **JAPAN**	HITACHI 5-1 Marunouchi 1-chome Chiyoda-ku Tokyo 100 (03) 212-1111
JVC **JAPAN**	JVC 1, 4-chome Nihonbashi Honcho Chuo-ku Tokyo 103 (03) 241 7811
MATSUSHITA **JAPAN**	MATSUSHITA 1006 Kadoma Kadoma City Osaka 571 Japan (06) 908 1121
PHILIPS **UK**	PHILIPS ELECTRONICS Professional LaserVision Department City House 420/430 London Road Croydon, Surrey CR9 3QR (01) 689 2166
PHILIPS **USA**	NORTH AMERICAN PHILIPS 100 East 42nd Street New York, New York 10017 (212) 697 3600
PHILIPS **HOLLAND**	PHILIPS EINDHOVEN Eindhoven, Holland (40) 79 1111
PIONEER **UK**	PIONEER HIGH FIDELITY Field Way Greenford, Middlesex UB6 8UZ (01) 575 5757

PIONEER USA	PIONEER VIDEO INC 200 West Grand Avenue Montvale, New Jersey 07645 (201) 573 1122
PIONEER JAPAN	PIONEER 4-1 Meguro-ku Tokyo 153 (03) 494 1111
RCA UK	RCA VIDEO DISCS 50 Curzon Street London W1Y 8EU (01) 499 4100
RCA USA	RCA 3600 North Sherman Drive Indianapolis, Indiana 46201 (317) 635 9000
SONY UK	SONY CORPORATION Pyrene House Sunbury-on-Thames TW16 7AT (76) 81211
SONY USA	SONY CORPORATION 9 West 57th Street New York, New York 10019 (212) 371 5800
SONY JAPAN	SONY CORPORATION 7-35 Kitashinagawa 6-chome Shinagawa-ku Tokyo

TECHNIDISC USA	TECHNIDISC Producers Colour Services 2250 Meijer Drive Troy, Michigan 48084 USA (313) 435 7430
THORN EMI UK	Thorn EMI Video Disc Metropolis House 39-45 Tottenham Court Road London W1P 9PD (01) 636 7694
VHD USA	VHD Corporation of America 4042 Blackfin Avenue Irvine, California 92714 (714) 660 9294
3M USA	3M OPTICAL RECORDING PROJECT 223-5S 3M Centre St Paul, Minnesota 55144 USA (612) 733 4758

SOME MAKERS AND DISTRIBUTORS OF INTERFACE PACKAGES, AUTHORING SYSTEMS AND DELIVERY SYSTEMS

ALLEN COMMUNICATIONS
7490 Clubhouse Road
Boulder, Colorado 80301

(303) 530 7300

AURORA SYSTEMS INC
2040 East Washington Avenue
Madison, Wisconsin 53704

(608) 249 5879

BCD ASSOCIATES INC
Suite 101
5809 SW 5th Street
Oklahoma City, Oklahoma 73128

(405) 948 1293

BELL & HOWELL AV LIMITED
Alperton House
Bridgewater Road
Wembley, England

(01) 903 5411

BELL & HOWELL
Interactive Communication Division
Marketing Services Department
7100 North McCormick Road
Chicago, Illinois 60645

(312) 673 3300

CAVRI SYSTEMS INC
26 Trumbell Street
New Haven, Connecticut 06511
USA

(203) 562 4979

CP SOFTWARE
21 Braybon Avenue
Brighton, England

(0273) 564500

FELIX LEARNING SYSTEMS LIMITED
25-27 Farringdon Road
London EC1M 3HA
England

(01) 404 5041

MILLS & ALLEN COMMUNICATIONS LTD
1-4 Langley Court
Long Acre
London WC2E 9JY
England

(01) 240 1307

OWL MICRO-COMMUNICATIONS
The Maltings
Station Road
Sawbridgeworth, Hertfordshire
England CM21 9LY

(0279) 723848

SCICON COMPUTER SERVICES LTD
Brick Close
Kiln Farm
Milton Keynes MK11 3EJ
England

(0908) 565656

INTERACTIVE VIDEO, VIDEODISC AND OPTICAL DISC NEWSLETTERS

INTERACTIVE VIDEO NEWS
EPIC Industrial Communications
28 Litchfield Street
London WC2H 9NJ
England

(01) 240 5863

NEBRASKA VIDEODISC DESIGN/PRODUCTION GROUP
PO Box 83111
Lincoln, Nebraska 68501
USA

(402) 472 3611

OPTICAL MEMORY NEWS
PO Box 14817
San Francisco, California 94114
USA

(415) 621 6620

SCREENDIGEST
37 Gower Street
London WC1E 6HH
England

THE VIDEODISC MONITOR
PO Box 26
Falls Church, Virginia 22046
USA

(703) 241 1799

VIDEODISC NEWS
PO Box 7005
Arlington, Virginia 22207
USA

(703) 241 1799

VIDEODISC NEWSLETTER
British Universities Film Council
55 Greek Street
London W1V 5IR
England

(01) 734 3687

VIDEODISC/VIDEOTEX
Meckler Publishing
520 Riverside Avenue
Westpoint, Connecticut 06880
and
3 Henrietta Street
London WC2E 8LU

INDEX